Metric (SI) in Everyday Science and Engineering

Stan Jakuba

Published by:
Society of Automotive Engineers, Inc.
400 Commonwealth Drive
Warrendale, PA 15096-0001

Library of Congress Cataloging-in-Publication Data

Jakuba, Stan
Metric (SI) in everyday science and engineering / Stan Jakuba.
p. cm.
Includes bibliographical references and index.
ISBN 1-56091-287-1
1. Metric system. 2. Units. I. Title.
QC91.J35 1992
530.8'1--dc20 92-30112
CIP

ISBN 1-56091-287-1

Preface

This book is intended for the person who needs a practical working knowledge of the units of the modernized metric system (SI). Undoubtedly, the text and figures will be useful also to those who have had some experience with the metric system and want to update their knowledge, and to all who are interested in overcoming the normal human tendency to dislike a new system that they do not understand.

The material is presented with the intent to implant the awareness of the simplifications possible with the new units, to warn of potential pitfalls associated with their use, and to guide in the recognition of which metric units and practices are correct and which are now obsolete.

Chapter 3 deals with physical quantities and their units — those needed in daily life and in engineering, particularly mechanical. Among them, units which are new to most Americans are covered thoroughly, while metric units already familiar (volt, ohm) are usually only listed in an appropriate section, although more is sometimes written to explain a specific detail. The supporting chapters deal with prefixes, rules for writing symbols and numbers, explanation of base and derived units, thoughts on converting and rounding, tables of reference numbers and conversion factors. The appendices present a brief history of metrication in this country and in the world, and an introduction to standards and standards-setting organizations with hints on how to change over to "metric."

The text and figures are organized for ease of orientation. The sequence of units in Chapter 3 follows the natural progression from the everyday topics to the specialized ones. Experience indicates that the resistance to using data and performing calculations in SI stems from the lack of a feel for the "ballpark figures." To provide some feel, most sections end with a listing of easy-to-remember items for the common sizes of the unit covered in that section. Most sections also include a short listing of rounded-off conversion factors.

The spelling of units and writing of symbols and numbers have been meticulously scrutinized for correctness. They reflect the most recent update of the rules.

Acknowledgements

This book is the result of unceasing expansion and modifications of this author's metric training material. Segments of the material were published over the years in various papers, articles, and Letters to the Editor. The lion's share of the reason for the existence of this complete text belongs to the thousands of students who sat through the training classes, asked the right questions, commented on the usefulness of the material for their needs, and pointed out the work's shortcomings and highlights. Without their contributions, approval and encouragement, the book would not appear in this form.

The author wishes to acknowledge also the advice and encouragement he received from the many proponents of the changeover to metric in this country, many of whom are authorities on SI. Among them, thanks belong foremost to the officers of the U.S. Metric Association, particularly Mrs. Valerie Antoine and Mr. Louis Sokol who were always ready with advice and an editing touch. The author is also indebted to Dr. Uri Gat of the Oak Ridge National Laboratories, and Mr. John Benedict, formerly of Chrysler Corp., for clarifying statements that could have had erroneous interpretation, and for helping to make the text shorter and more concise.

Further, he is grateful to Dr. Barry Taylor of NIST for providing updates on the latest, as yet unpublished developments in the standards on units. The author also most gratefully acknowledges the patience and faith of his wife Eva who never tired of listening and contributing to the discussions about metric, and the willingness of his children to use this material in school and speak SI at home.

Finally, he is grateful to SAE for publishing the work.

Table of Contents

Chapter 1
Introduction

The international system of units, known as SI in all languages, is the only complete, state-of-the-art, maintained and recognized system of units. Based on the "metric system," SI was created in 1960 and is now replacing all other systems of units for use worldwide. This includes the U.S. and Imperial versions of the so-called English or inch-pound system, and the previous SI versions of the metric system (CGS, MKS, etc.). All nations, metric as well as non-metric, are in the process of adopting SI. The universal use of SI is the goal that, though not yet attained, is becoming a reality.

Background

The metric system of units, called metric after one of its units which was named from the Greek word for a measure, began its life during the French revolution (1791) as an attempt to standardize measures. It was conceived by scientists, most of them French, who devised the few fundamental units upon which the system later developed. Interestingly, the first country to retain the new units was Switzerland, not France.

Why did the effort for a unified system of units succeed at that time? After all, this was not the first time such an attempt was made.

There existed, as we know, many "systems" of units in history. Within a system any given unit may have had a different size from one city to another, from one generation to another. Unified systems of units were proposed many times, but it was not until the emergence of industry and the opportunity offered by the revolution that made it both necessary and possible to introduce such unification on a wide scale.

From a 19th-century perspective, the spread of the metric units through Europe was quite rapid, propelled undoubtedly by French troops trading in the occupied lands during the Napoleonic wars. Neighboring European nations followed suit as industrialization forced them to further their commercial horizons.

Similarly, in other parts of the world, native systems were replaced by those of the colonizing powers or those of a more influential trading partner. By the end of the century only nine "systems" remained. The natural process of elimination removed all but two — the so-called English or inch-pound system, and metric (these are shown below). Both have been modified several times and are concurrently used in more than one version.

Two Measuring Systems

Inch-Pound (English)

Imperial
U.S. (Pre-Imperial)

Metric

CGS (Absolute)
Technical (Gravimetric)
MKSA

As is known, this country uses the U.S. version of the inch-pound system, while the other English-speaking countries had used the Imperial version. Although named identically, several units were of different sizes in the respective versions (recall the sizes of the U.S. and the Canadian gallon of gasoline).

The differences among the versions of the metric system were not as pronounced. For all practical purposes, the size of units never changed; the changes redefined units and terms, added new ones, eliminated others. Some of the obsolete metric units and terms are still used and may have different meaning among localities. Thus, a person trying to learn "metrics" by studying documents which contain metric data often faces a confusing ordeal.

Help is coming. By now practically all nations have imposed a mandate requiring that only SI be used, and they are working on implementing it. The world is in transition to SI. During this time, some countries, companies, or even departments within an institution will inevitably differ in their interpretation of regulations and use "old metric," SI, or a mixture of all in their documents. Older metric data should therefore be examined carefully and modified to be SI for future use.

The existence of the non-SI versions of the metric system obscures the inherent simplicity of SI, and makes it seem that learning metric is difficult. It is not so when learning SI.

Features of SI

As in all versions of the metric system, the sizing of units in SI is based on decimal arithmetic. Why decimal? Because, throughout history, mankind most often preferred counting with tens, and this preference is undoubtedly based on the fact that people have ten fingers.

SI is unique among the other versions of the metric system for its strict and practical coherence. It has only one unit for any physical quantity and therefore no conversion factors and constants to remember.

Note on coherence: A coherent system of units is simple. In practical terms coherence means that, if, for example, the inch-pound system were made coherent, it would have only one unit of length. That unit and its derivatives for area and volume would replace all other length and length-derived units, such as angstrom, mil, foot, fathom, mile, acre, pint, gallon, bushel and barrel. Conversely, instead of the ounce which in the present inch-pound system denotes two different volumes, two different masses and a force, a coherent system would have an individual unit for each of the three quantities. All units describing independent quantities would be independent (arbitrary, base), and all other units would be derived from them in the same fashion that the physical quantities they describe derive from the base properties.

Learning SI

The beauty of the coherence and decimal arithmetic which make SI so simple is apparent throughout this material. In comparison to other systems, SI is the easiest one to learn.

SI is easier to learn than the inch-pound system; but because the material presented here on a few pages is normally learned in many years while growing up, the learning does require some work. The fact that it can be done at all in such a short span is partially due to the simplicity inherent in SI, and partially because many metric units have been used in the U.S. for generations.

FAMILIAR "METRIC" ITEMS

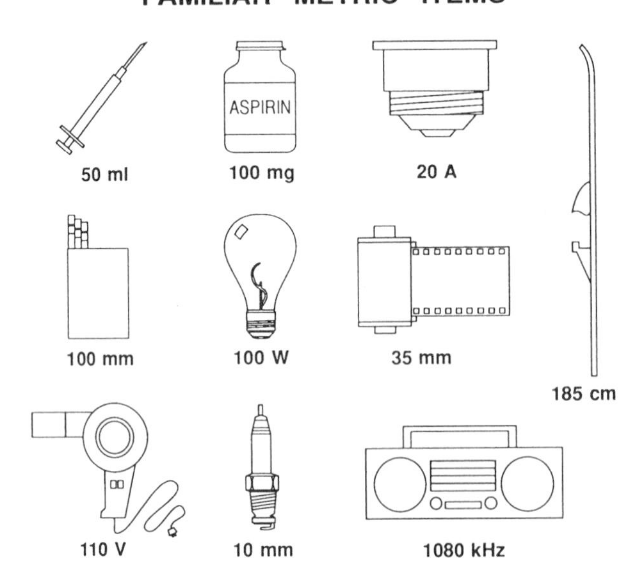

As a matter of fact, most of the hundreds of units used in the U.S. today are metric; only a few dozen are not. Unfortunately, those few are the units of the most commonly used quantities such as length, area, volume, mass, and temperature.

The following chapters describe features of SI. The material is presented from the practical point of view, and it includes considerations for communicating technical information among different nations and languages.

When an engineer or technician or any person understands the principles and reasons designed into SI, that person becomes better equipped to solve problems, and is thus that much more valuable to his or her profession and society. This book will help you with that understanding.

Chapter 2
Prefixes

Prefixes are symbolic names (kilo, mega, milli) that often precede the name of a unit.

Note: The reader completely unfamiliar with prefixes is advised to read the first section in Chapter 3, Physical Quantities and Their Units, before continuing. Prefixes and the system behind them are easier to understand if an application is observed first.

Prefixes were devised to shorten lengthy numbers, an alternative to the 10^n notation (e.g., 5000 m shortened to 5×10^3 m) and to the creation of new names (e.g., 3520 yards shortened to 2 miles). The 10^n notation is impractical for the non-scientific person, while the creation of the new names is impractical for everyone, as it would necessitate the creation of hundreds of new names.

While SI currently recognizes 20 prefixes, only four prefixes are usually needed in common life, and another four in common engineering. These eight prefixes are shown below; a listing of the others is at the end of this chapter.

Review of Prefixes

Value	Prefix Name	Prefix Symbol	Example
10^9 (billion)	**giga**	**G**	**GJ (gigajoule)**
10^6 (million)	**mega**	**M**	**MW (megawatt)**
10^3 (thousand)	**kilo**	**k**	**kg (kilogram)**
10^0 (one)	**-**	**-**	**V (volt)**
10^{-1} (tenth)	deci	d	dm^3
10^{-2} (hundredth)	centi	c	cm^3
10^{-3} (thousandth)	**milli**	**m**	**mV (millivolt)**
10^{-6} (millionth)	**micro**	**µ**	**µA (microampere)**
10^{-9} (billionth)	**nano**	**n**	**ns (nanosecond)**

Notice that:

— The prefixes printed in bold differentiate the size by a factor of 1000. (The three-digit differentiation is both preferred in SI and always used in engineering.)

— The prefixes printed in regular type (deci and centi) do not fit the 1000 differentiation. Notice, however, that when used with the cubic measures, as shown in the examples, each does represent a thousand-fold change. (Deci and centi used to be common with several units in the past, but they are nowadays used by the public with the unit of length, area and volume only, and engineers prefer to use them only with volume.)

— The symbols of prefixes indicating a value larger than k use an uppercase letter and those indicating a smaller value use a lowercase letter.

The accent in pronunciation of a name with a prefix is on the first syllable, just as if the prefix were pronounced alone. For example, we say **ki**logram, not ki**lo**gram.

Prefixes must not be combined — there can be only one prefix with a unit. For example: 1 Gg, not 1 Mkg.

Any prefix is usable with any unit. Convention, however, restricts the usage of some prefixes in certain applications, as, for example, the above-mentioned use of deci and centi. And as with all SI symbols, the symbols of prefixes are the same in all countries and languages; they are not abbreviations.

Note: Prefixes that are needed in a profession, and the value each represents should be memorized. This is best done by learning the sequence. The effort of memorization is small in comparison to the advantages the knowledge brings.

A prefix tells how many times larger or smaller a unit is. This feature is not available in the inch-pound system; the word mile, for example, bears no clue as to how it relates to other units of length, such as the yard.

Prefixes also make any unit comfortably sized. With them, no SI unit is "too small" or "too large." A few seconds of concentrated effort once in a lifetime is all that is needed to learn how to become comfortable with the size of any unit in any discipline.

For reference, the other prefixes are:

yotta (Y) = 10^{24}	hecto (h) = 10^{2}	pico (p) = 10^{-12}
zetta (Z) = 10^{21}	deka (da) = 10^{1}	femto (f) = 10^{-15}
exa (E) = 10^{18}		atto (a) = 10^{-18}
peta (P) = 10^{15}		zepto (z) = 10^{-21}
tera (T) = 10^{12}		yocto (y) = 10^{-24}

The prefixes hecto and deka are fast becoming obsolete. Notice that they do not fit the 1000-increment preference.

Note: The worldwide spelling of the sound "k" and the pronunciation of some prefixes:

Worldwide, one often sees "k" in hekto, deka, mikro, and also piko. The first three reflect the original Greek spelling. In this country, the "c" spelling is preferred, except in deka. More about spelling is at the end of the section on the unit of length in Chapter 3, and in Chapter 5, Rules For Writing Units and Numbers in SI.

The pronunciation of the prefixes giga and micro also differs among languages. The phonetic pronunciation, however, seems to prevail worldwide. Thus both "g's" in giga tend to be pronounced the same (as in gargantuan), and the "i" in micro sounds as in milli, giga, kilo, or pico.

Chapter 3

Physical Quantities and Their Units

The sections in this chapter are arranged to approximate the frequency with which the physical quantities are encountered in daily life, the common ones first. However, where advantageous for pointing out the logic or relationship of units, the not-so-common quantities may be presented earlier.

The quantities are listed in the following order:

Length, Area, Volume, Section Modulus, Moments of Area
Mass
Time
Temperature
Angle
Force and the Weight/Mass Distinction
Energy and Torque
Power
Pressure
Frequency
Acceleration
Vibration
Other Quantities Used in Longitudinal Mechanics
Density and the SI Terms for "Specific Weight" and "Specific Gravity"
Viscosity
Moment of Inertia and Other Moments
Other Quantities Used in Rotational Mechanics
Amount of Substance
Electricity
Light
Radiology

Note on semantics: Throughout this text, the term "unit" is used interchangeably, having any one of these meanings:

— a name of a unit (e.g., watt)
— a symbol of a unit (e.g., W)
— a name of a unit multiple (e.g., megawatt)

— a symbol of a unit multiple (e.g., MW)
— a name of a unit submultiple (e.g., milliwatt)
— a symbol of a unit submultiple (e.g., mW)

Length and Length-Associated Quantities: Length, Area, Volume, Section Modulus, Moments of Area

Length, Area, Volume

The unit of length is the meter (metre, metr, metro, μετρ; for spelling see the footnote at the end of this section). Its symbol is m.

This unit and its two derivatives, the unit of area (m^2) and the unit of volume (m^3), serve in places where the inch-pound system retains a whole array of units, such as foot, inch, acre, cubic yard, fl. ounce, gallon, bushel, hogshead, and barrel.

The figure below shows how m, m^2, and m^3, and their submultiples, relate to parts of the human body and to some common objects.

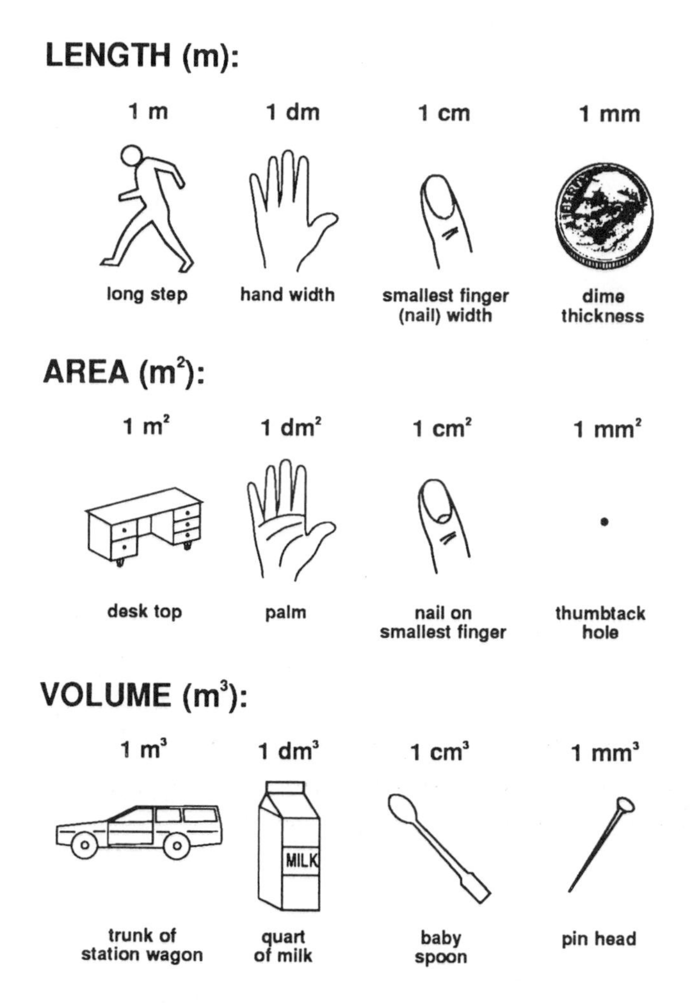

The size represented by the symbol dm^3 is commonly referred to as the litre (liter, litr, ..). The symbol cm^3 is often denoted by mL (ml), and, incorrectly, in this country also by cc. In engineering, and in all technical writing, only the SI symbols (.., km^3, m^3, dm^3, cm^3, mm^3, ..) should be used.

In engineering drafting (in any engineering and in any part of the world), dimensions on drawings are always in mm only. This rule is often observed in other technical documents; they show dimensions in numbers without a unit because mm is understood.

Also, engineers shun the prefixes deci and centi except with volume, as was mentioned in Chapter 2, Prefixes. For engineering purposes, therefore, the earlier picture changes in the Length and Area section as shown below.

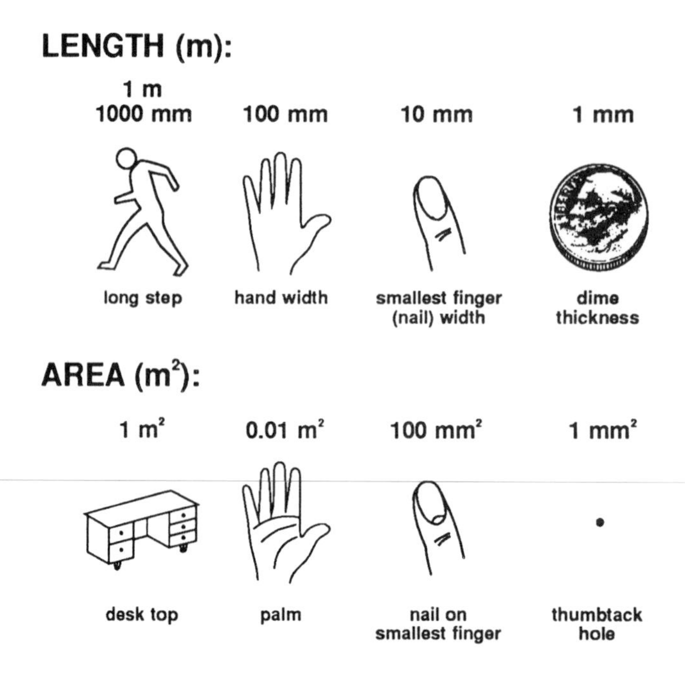

Notice that the prefixes d and c are replaced with the prefix m or with no prefix at all, and the numbers are adjusted accordingly.

In daily life, on the other hand, commodities such as furniture and fabric are usually sized in cm, as are measurements in carpentry and sporting events. Similarly, dm is used in forestry and in ocean and waterways charting.

Section Modulus, First and Second Moment of Area

The units of section modulus and the first and second moment of area are also derived from m; they are the m^3, m^3 and m^4, respectively. More about these quantities is in the discussion on Moments.

A Feel For Sizes

Length:	1 µm: hundredth of paper thickness
	1 mm: thickness of a dime
	1 m: length of a man's long step
	1 km: length of ten football fields
Area:	1 mm^2: large paper clip wire cross-section
	1 m^2: office desk top
Volume:	1 cm^3: baby spoon; fifth of teaspoon
	1 dm^3: quart of milk (filled to the "brim")
	1 m^3: trunk of a mid-size station wagon

Equivalent Common Values

Length:	1 yard is a bit short of 1 m
	1 ft is a bit short of 0.3 m
	1 in. is exactly 25.4 mm
Area:	1 in^2 is just under 650 mm^2
	1 ft^2 is a bit less than 0.1 m^2
	1 acre is just over 4000 m^2
Volume:	1 quart is a bit short of 1 dm^3
	1 gallon is a bit less than 4 dm^3

Footnote: Spelling of the Unit of Length

The spellings for written-out SI units are not unified; they may be written according to the local language and script as is illustrated by the words in the parentheses. In the English-speaking countries outside the U.S., the metre spelling is used. In the U.S. both the metre and meter spelling is used; the preference differs from one manual to another and sometimes from one issue of the same manual to another.

To avoid the -re, -er controversy, only the symbol of the unit will be used in the further text. SI symbols are identical in all languages regardless of the script of a language. The symbol for the unit of length looks, to the person familiar with the Latin script, like the lowercase m. Symbols are not abbreviations.

Mass

Two topics are covered in this section: (1) the unit — its origin and the complication of the prefix in its name, and (2) the change in the terminology of the word "weight."

The unit (kilogram):

When the founders of the metric system thought about sizing the unit of mass, they decided to fix it on the quantity of water filling a certain metric volume. They selected the volume of 1 cm^3; the mass of water it contained became the gram, symbol g.

So it happens that 1 m^3 contains 1 Mg of water, 1 dm^3 contains 1 kg, 1 cm^3 contains 1 g, and so on. This relationship between the volume and mass of water is a handy and convenient feature, most useful both in daily life and in engineering.

The figure below shows how the size of the gram and its multiples and a submultiple relate to some common objects. Remembering these pictures, and remembering the relationship between mass and volume of water makes developing a feel for the unit of mass easy indeed.

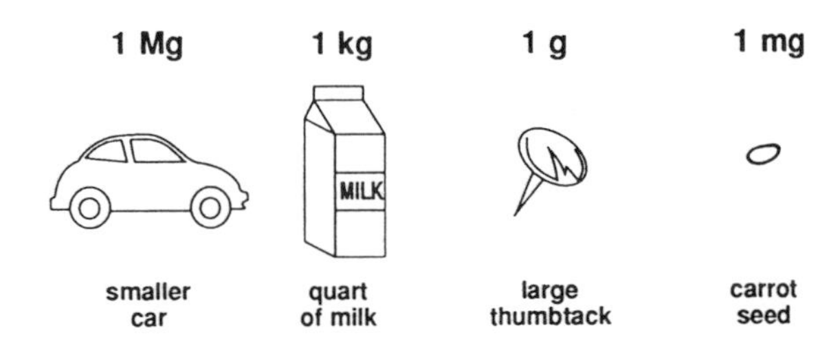

The size represented by the symbol Mg is often called the metric ton (tonne). Because the word "ton" has several different meanings and spellings, its use is discouraged. In technical writing, the SI symbol should always be used.

In engineering and scientific calculations, it is important to consider that the kilogram, not the gram, is the base unit of mass in SI (the term "base unit" is described in the section of this chapter titled Base and Derived Units). The kilogram figures in formulas which involve derived units such as newton, joule, and pascal (see later sections).

Note: The thousand-fold increase in the size of the unit of mass was made with one of the earlier revisions of the metric system. A new name was not coined, and thus today the unit of mass, the kilogram, has a prefix in its name. Since adding a prefix to a prefix could cause confusion, multiples and submultiples of the unit are formed with the word gram.

The terminology change:

Mass is the correct word to describe the amount of material, the property that gives an object its inertia, the quantity we measure with the above-mentioned unit.

The word "weight" is no longer to be used for this quantity nor should weight be associated with the unit of mass. The meaning of weight and the reasons for this are described in the section titled Force and the Weight/Mass Distinction. The methods for measuring mass are also shown there.

The term "mass density" is discussed in the section titled Density.

Note: In some versions of the metric system, as well as in the inch-pound system, grams, pounds, etc., were also the units of force. This is not the case in SI.

A Feel For Sizes

1 mg: dry carrot seed
1 g: dollar bill; large thumbtack; large paper clip
1 kg: quart of milk (with the carton)
1 Mg: smaller car; ten big men

Equivalent Common Values

1 carat (metric) is 0.2 g
1 oz is about 30 g
1 lb is a bit less than 0.5 kg
1 long ton is about 1 Mg

Time

The second is the SI unit of time. Its symbol is s (not sec). As there is only one unit for any physical quantity in SI, the second replaces all the other time units, such as minutes and hours.

In engineering, time usually means an interval, as when measuring flow and velocity. Here the second introduces simplicity, makes for easier comparison, and should be used exclusively; engineering stopwatches should count to hundreds of seconds, not change to minutes.

In daily life, however, where time usually means a point on a scale, the traditional minutes and hours are retained as they have been universal throughout the world. The only remaining step here is to unify on the 24-hour clock dial.

Note on the "decimal hour": There have been numerous attempts in history to divide the day decimally. Although at times successful in some societies, the attempts have not succeeded in modern history except for brief periods. Strictly speaking, this is not a subject pertinent to SI.

Equivalent Common Values

1 min is 60 s
1/4 h is a bit shorter than 1 ks
1 h is exactly 3.6 ks

Temperature

The kelvin is the SI unit of temperature. Its symbol is K (not °K).

For the purpose of obtaining a feel for the size of a kelvin, the kelvin is one hundredth of the difference between the freezing and boiling temperatures of water.

Most people are familiar with the degree Celsius, symbol °C, which is a special name that may be used in place of kelvin. There is no difference between a Celsius degree and a kelvin in terms of a temperature increment. On the other hand, when referring to a point on a temperature scale, 0 K is absolute zero, and 0 °C corresponds to approximately 273 K.

Notice the two meanings here: an increment, and a point on a scale. Accordingly,

- as an increment, 1.8 °F ≡ 1 K ≡ 1 °C ≡ 1.8 °R;
- as a temperature level, 1.8 °F ≡ 256.4 K ≡ -16.8 °C ≡ 461.5 °R.

The double meaning makes a difference in the selection of the proper conversion factor. Mistakes are frequent when converting °C and °F; they are less likely with K and °F. A case example: A tolerance of 10 °C was mistakenly converted to 50 °F instead of 18 °F. In converting K and °F this is unlikely to happen as the large difference makes the mistake obvious.

The kelvin, an increment or a scale, is preferred in engineering calculations, standards writing, low- and high-temperature work, and is gaining acceptance in other fields. It is recommended for use always when referring to an increment. The degree Celsius, on the other hand, is firmly ingrained in the public. Here the concern is almost always for temperatures of the weather range; for this the Celsius scale is hard to give up.

In this interim, a specification in the form of 27 °C ± 2 K is acceptable if the preferred form 300 K ± 2 K, or (300 ± 2) K, cannot be used. The unit best accompanies each number, or parentheses are used as shown here to avoid misinterpretation.

The following figures show conversion formulas, a scale with several melting points, and the values of typical weather temperatures in kelvin.

KELVIN SCALE CONVERSION:

From °F: Temperature in K = (temp. in °F + 459.67)/1.8
Example: 80 °F = 299.82 K ≐ 300 K

From °C: Temperature in K = temperature in °C + 273.15
Example: 27 °C = 300.15 K ≐ 300 K

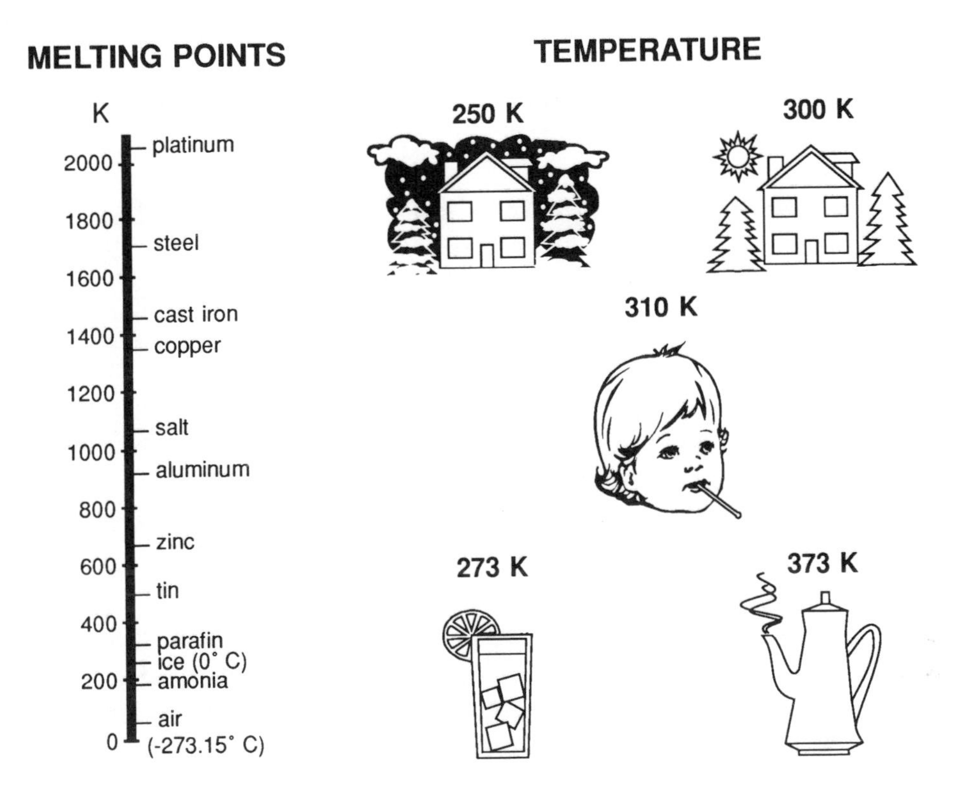

Equivalent Common Values

As an increment: 1.8 °F is exactly 1 K
As a scale: 0 °R is exactly 0 K
0 °C is a bit above 273 K

Remark: Base and Derived Units

Base Units:

Among the units introduced so far were m, kg, s, and K. These four units and the units A (ampere), cd (candela) and mol (mole) are called *base* units.

Ampere is the unit of electric current, and candela is the unit of luminous intensity. The usage of these two units has changed insignificantly with the establishment of SI and they are therefore not dealt with further except for their listing in the sections on electric units and light units. Mole is the unit of amount of substance. Its definition and the related terminology are covered in the section titled Amount of Substance.

These seven units can be thought of as the foundation of SI. Their size is arbitrary.

Derived Units:

All units other than the base ones are called *derived* units. There are hundreds of them, and new ones are added continuously to fulfill the needs of an evolving civilization. They are formed by an arithmetic combination of base units, a combination of other derived units, or a combination of units from both groups.

The preferred way of writing a derived unit is by placing between the symbols a raised dot for multiplication (e.g., g·m) and a slash for division (e.g., m/s). This is covered in more detail in Chapter 5, Rules for Writing Units and Symbols in SI.

A small number of derived units have been given special names. This was done to shorten the writing and pronunciation of the base-units-derived symbols (e.g., N for $kg{\cdot}m/s^2$), or to distinguish units that might otherwise look alike (e.g., 1/s and Hz). Many of the specially named units are familiar, particularly the units related to electricity (volt, farad, ohm).

The next sections deal with the derived units frequently used in engineering and daily life that are either new with SI or have been used already but often pose difficulties. Two derived units, the unit of area and the unit of volume, were discussed earlier.

Note on the terms Supplementary Units and Compounded (Composite) Units:
Literature often lists radian and steradian as supplementary units. The current international agreement effectively puts these two units into the derived unit category, and that is the guidance adhered to here. Radian and steradian are *dimensionless derived* units.

When a derived unit contains both a derived unit with a special name and a base unit or other derived unit, literature sometimes lists it as *compounded*, or *composite derived* unit.

Angle

There are two angular measures: a plane angle and a space (solid) angle. The unit of the former is the radian, symbol rad; the unit of the latter is the steradian, symbol sr.

One steradian is the solid angle that cuts off an area on the surface of a sphere equal to the square of its radius.

One radian is the plane angle subtended by the length of the circumference equal to the radius. This relationship is shown in the following figure.

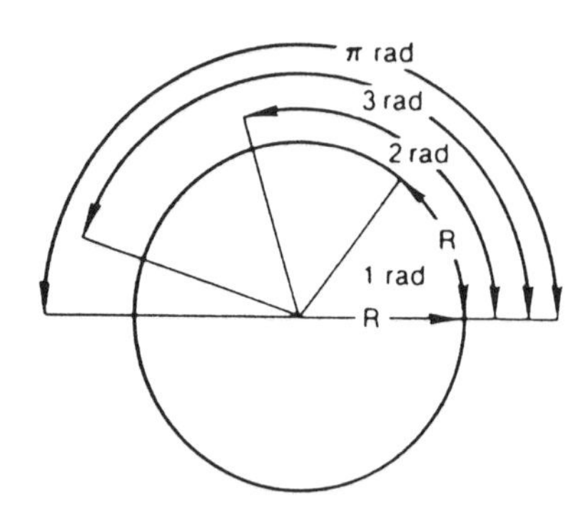

It is apparent from the figure that the "dimension" of radian is 1 (or rad^0), as it is a ratio of two lengths. Consequently, it is permissible to omit radian when no confusion can result. For example, to express the angle of $\pi/3$, rad need not accompany the number, while to express the unit of torsional elasticity, rad must be stated to avoid ambiguity.

The radian is a convenient unit for many engineering calculations and has been used there for generations. On the other hand, the radian is considered inconvenient for drafting, carpentry, surveying, etc. Therefore, although not coherent, the traditional degrees, ratios and percents are permitted to continue. Only the use of minutes and seconds is discouraged; they should be replaced with decimal divisions of a degree. As for the ratio in slopes, cones and tapers, it is always formed by two identical units (e.g., m/m, or mm/mm, never mm/m, in/ft, etc.).

Example: 22°30′ should be either 22.5°, or 0.125π, or $\pi/8$, or 0.39 rad, or 41%, or 1:2.5 (the last three expressions were rounded off).

Because of the multitude of units, toleranced angles should be written with units at both numbers or with parentheses. Examples: 20° ± 0.5°; (20 ± 0.5)%.

Equivalent Common Values

30 degrees is $\pi/6$ rad
57.3 degrees is about 1 rad
60 degrees is $\pi/3$ rad
100 grad (gon) is $\pi/2$ rad
full circle is 2π rad

Force and the Weight/Mass Distinction

The unit of force is the newton, symbol N.

The force of 1 N feels like the gravity pull of a small apple. Recalling the story about Isaac Newton under the apple tree helps to retain the feel for force in newtons.

The unit newton is defined as the force needed to act on a 1 kg mass to accelerate it at 1 m/s^2. Thus, the unit of force in base units is $kg{\cdot}m/s^2$. An expression inconveniently complex, it was given the symbol N. Recalling Newton's second law: *force = mass × acceleration*, helps one to remember the relationship between N and $kg{\cdot}m/s^2$.

A mistake-free application of the unit newton requires that the terminology of the word "weight" be clarified first. Not everyone realizes that "weight" is ambiguous. The word has at least three meanings in many languages. People in non-metric as well as in metric countries are learning to distinguish the circumstances when "weight" should be assigned the mass unit and when the force unit. Understanding this issue helps also in assigning the proper unit to "load" and similar terms.

The following text is intended to explain the issue and thereby enable the reader to select the correct unit, conversion factor, and application in all cases.

Note: To keep the text brief, the word gravity will often be used to mean what is known as acceleration of gravity, acceleration of free fall, and similar terms.

The Two Meanings of "Weight"

Two of the usual meanings of weight are of interest to us here. They are analogous to the two sensations a person has when handling an object.

1) The one sensation is recognized as the force the person exerts holding the object. In this context, weight refers to *gravity force*, and it is the net force that pulls the object downward (there are other definitions further in the text).

2) The second sensation is recognized as the inertia of the object. In this context, weight refers to *mass*. This sensation exists regardless of the presence or absence of gravity; in a "gravity-free" environment, the mass of an object is apparent when the object is accelerated or decelerated, including when the direction of its motion is changed.

The international body (CGPM) governing the development of the international system of units declared the word "weight" to mean force, i.e., the first sensation.

We shall, however, avoid the word "weight" altogether in this text (except where needed to illustrate a point). It is easier to explain the subject when "weight" is consistently replaced by the term of the intended meaning, which is mass or gravity force here. The usage of the term "mass" reflects the ruling presented in the section on Mass. The usage of the term "gravity force" parallels the common practice of forming two-word terms such as centripetal force, exciting force, and magnetic force.

How to Recognize When to Consider Gravity Force and When to Consider Mass

We have defined gravity force as the net downward force acting on a stationary object. It can also be defined as the force needed to support an object. The force is always proportional to the strength of the gravitational field and the object's mass.

On Earth, the strength of the field is fairly constant along the surface, and therefore an object is pulled downward with very much the same force regardless of where it is. Thus, seemingly, gravity force and mass of an object are interchangeable; their numerical values and their units could be the same. Indeed both quantities have in daily life been treated as such: they have shared one word (weight) and have had identical units (lb, oz, kg, t, ..).

Many measuring systems (of which there were dozens in history) assigned a different unit to mass and force, but the distinct units remained confined within scientific circles. When the need was recognized in engineering, modifications of the existing units were introduced: subscripts (lb_m, lb_f, oz_m, etc.) in the inch-pound countries, and subscripts (g_f), asterisks (kg*), the word "pond" (kp), etc., in the metric countries. Except for some diligent students and pedantic professors, the modifications were usually ignored.

SI prescribes a clearly distinct unit to each quantity: newton to force, gram (kilogram) to mass. Unlike the previous attempts, the SI approach is finding acceptance among the scientific community as well as among engineers and the general population.

The following examples illustrate the distinction.

Examples:

Example (1): The first example discusses gravity on Earth. As we know, the Earth's gravity pull is stronger at the poles than at the equator, and stronger at sea level than at elevations or depressions. An object is pulled downward more at the poles than at other locations.

Note: The reason for the uneven pull is the same as that which makes the globe oblate: the spinning on the pole-to-pole axis. The spinning causes an apparent centrifugal force that counteracts gravitational attraction. The resulting latitudinal gradient is, however, not directly related to the radius; the proportion is skewed due to the influence of the greater gravity mass in the plane closer to the equator, and it is also influenced by the local density of the Earth's crust.

Even at its most extreme, however, gravity varies by less than one percent among commonly accessible localities on Earth, so it matters little to most people. The variation is, of course, significant for scientific measurements, and it can also be important for other purposes, as this fictitious situation illustrates:

A country at a pole sends a certain quantity of gold to a country at the equator. A spring scale is used to measure the gold, and the scale is also sent. After delivery, the gold is found to be lighter. This despite the fact that it is measured on the same scale as before. Is a part of the gold missing? Let's look at this picture.

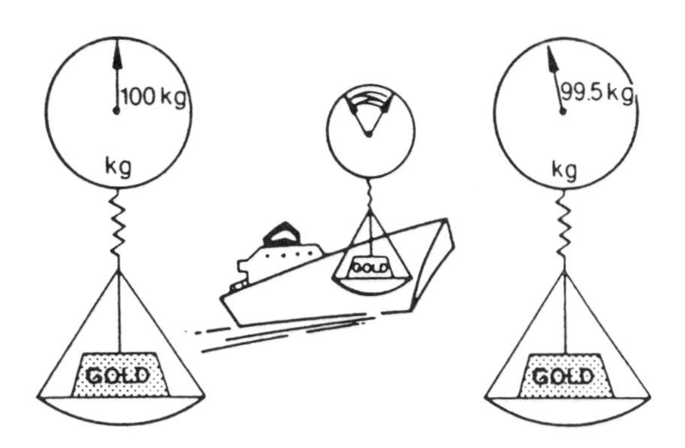

No part of the gold is missing, of course. The perceived disappearance was caused by the use of the spring scale. A spring scale senses a force, in this case the force with which the gold is pulled downward, i.e., the gravity force.

Gravity force is determined by both gravity and mass. If either or both change, gravity force changes too. Because gravity is reduced by the apparent centrifugal force at the equator, the scale reads less there; no part of the gold needs to disappear to get that reading.

How could this scale be modified to give the correct mass reading in the new location? It should have either narrower divisions on its dial, or a weaker spring. Mass can be measured directly by measuring gravity force only if the scale is built and calibrated for the local gravity.

The use of the term "weight" and the weighing procedure led to an error. The procedure was intended to measure the mass of gold but it measured its gravity force. The error is obvious if the unambiguous terminology is applied.

Notice that if the scale dial had displayed a force unit, the mass could have been calculated as shown below, and found to be correct.

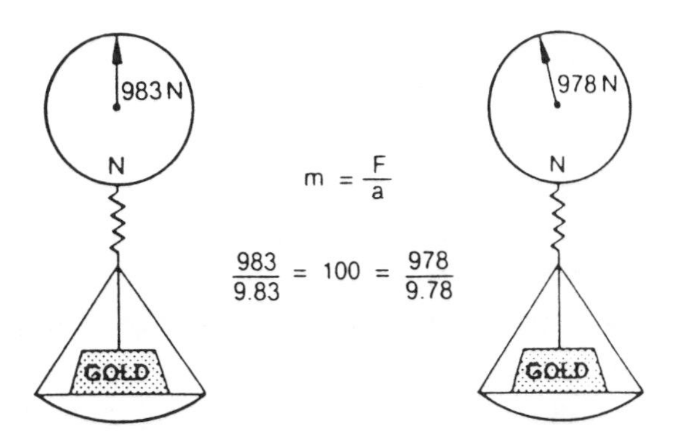

The choice of a term that is unambiguous must be made to fit circumstances, true with any material, except that with gold the circumstances are rather limited. Mass is almost always meant with gold; mass is meant when gold is traded, prices quoted, jewelry appraised. The gravity force is seldom of concern as the quantity of gold anyone carries around does not require an appreciable physical effort. (An example of the occasion when gravity force would be of a concern is in the calculation of the stress to which the elevator at a gold repository is subjected.)

To explain the distinction further, let's move away from Earth. The gravity force of an object on the Moon is one-sixth of the force exerted on Earth. In an orbiting satellite, gravity force of an object is zero because the attraction of the Earth and other heavenly bodies and the apparent centrifugal force of the orbit are in equilibrium. By contrast, the *mass* of an object is not changed by relocation, Earthly factors, or, indeed, by anything at all (in this context). And it is understood that the object can be a solid, liquid, gas, or a particle.

There are more factors that influence gravity force than presented in the first example. The next example is a listing of some of the others.

Example (2): As stated before, gravity force is the net effect of the various static, up-and-down forces acting on an object at a given location and time. It is the net effect of the Earth and object attraction at a particular locality, apparent centrifugal force, air buoyancy, attraction with the Moon and other heavenly bodies, and a few lesser forces.

Therefore, in addition to the previously mentioned location influence, gravity force changes with air pressure and temperature, position of Moon and Sun, etc. The changes are negligible for most engineering purposes, but matter in the calibration of precision instruments and some scientific experimentation.

Example (3): The last example illustrates the distinction by considering buying food in a grocery store. Modern grocery scales determine the mass of food by measuring the force by which it pushes onto the scale platform due to gravity. The scale is a *forcemeter*, just like a spring scale. (Forcemeters normally measure by an elastic deformation of a solid or fluid.)

Regardless of how the grocery is measured, our body needs the mass of the food to live, and mass is what we want to pay for. For that reason the scale dial is graduated in grams, not newtons. Obviously, the word "weight" implies mass here. From the foregoing it is apparent that one might pay the wrong amount if the scale were moved to a different altitude, latitude, etc.; grocery scales are, of course, seldom relocated far enough away to register a difference.

Meanings of "Load"

The distinction of meaning, similar to the one applied to "weight," now applies also to the word "load." This term, too, may have a different unit under different circumstances as it is commonly used to mean mass and gravity force (and also power, current, and others). As examples, consider the railroads: the load on a locomotive coupling is the tractive effort, clearly a force (N); the load the axle supports is gravity force (N); the load the freight cars transport is mass (g).

Measuring Mass or Gravity Force

Mass is most often measured by sensing the force the object exerts downward, i.e., by measuring its gravity force. The instrument is a *forcemeter*, the dial of which has been inscribed to display mass units as, for example, a common bathroom scale does. Such measurement is influenced by the previously mentioned factors.

To eliminate these influences, an object's mass is best measured by a comparison to a known mass. A scale for the direct measurement of mass is the balance scale, or *massmeter*, an example of which is shown here.

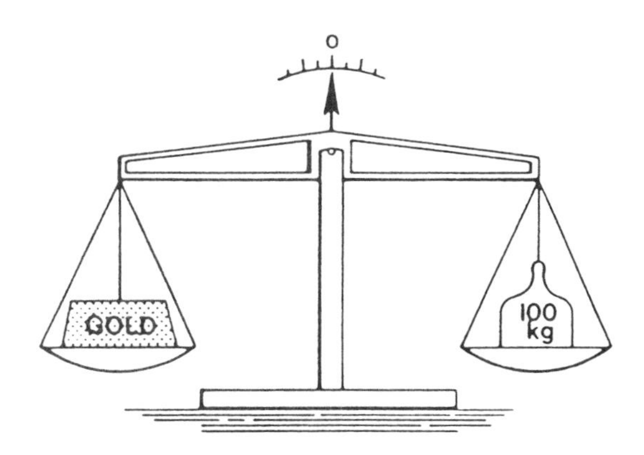

The mass comparison method is used for precious metal trading, for calibrations, and similar cases where the inconvenience of a massmeter is less important than its insensitivity to changes in gravity, its lower sensitivity to buoyancy changes, etc.

Although measurements using a balance scale are not influenced by changes in gravity, the scale functioning depends on the presence of some gravity level. Where there is no gravity force at all, as in an orbiting space station, mass could be determined by calculation from a known force and the resulting acceleration: $m = F/a$. In practice this is done by attaching the object to a spring system that, when disturbed, enables the mass to be calculated from the cycle time of the free oscillation.

Among the other methods used for determining mass the most common ones are the measurement by volume calibrated for a liquid of known density, and the calculation from volume, temperature and pressure for gases.

Gravity force is usually measured just like mass — on a forcemeter, the dial of which displays newtons, however. It can, of course, be determined by calculation from any known mass (or mass density) and local gravity.

Accuracy of Measurement

Forcemeters measure an object's gravity force accurately (in this context) anywhere, but they measure an object's mass accurately only under the gravity for which they were built. The graduation of their dial can be in newtons or grams to suit a purpose. The "newton" dial will be accurate everywhere, the "gram" dial will not.

Many other measuring devices sense one parameter but display another. An example of this is the car odometer: It senses the revolutions of a wheel but displays distance traveled. As the tire diameter changes with wear or speed, the distance reading becomes inaccurate. Similarly, if gravity changes, the reading of mass on a forcemeter will be inaccurate.

How does this agree with SI principles? The scale dial of a forcemeter displays the unit most suitable for a purpose; whether it is graduated in newtons or grams is not a question of being SI or not.

In Summary

In summary, the unit of any force is newton; force is defined according to Newton's law, *force = mass × acceleration*; hence, N is expressed in base units as $kg{\cdot}m/s^2$.

Note: Notice that the size of a newton is derived from kilogram, not gram. This is a consequence of the decision discussed in the section titled Mass. Engineers often have difficulties relating derived units to base units because of the 1000 factor.

Referring to gravity force, which is the force needed to support a stationary object, *gravity force = mass × gravity acceleration* ($F_g = m \times a_g$). On Earth, the acceleration ranges between 9.75 m/s^2 and 9.83 m/s^2. The values are such that, for rough estimates, a_g can be considered equal to 10 m/s^2. Rounding off leads to the simple rule that the numerical value of an object's gravity force (in N) on Earth is about ten times the numerical value of its mass (in kg).

Note: In some metric (non-SI) documents, kilogram, gram, etc., are still used as force units. Here the unit is usually written as kg_f or kg* or kp. To convert to SI, 1 $kg_f \equiv$ 9.806 65 N.

This conversion factor is a reference value. It means that, for reference purposes, gravity force equals 9.806 65 newtons per kilogram. Values for other purposes are presented in the section titled Acceleration.

Two weight-related quantities, Specific Weight and Specific Gravity, are discussed in the section on Density.

A Feel For Sizes

1 N: gravity force of a small apple (on Earth)
1 kN: gravity force of a heavy person (on Earth)
1 MN: thrust of jumbo jet engines

Equivalent Common Values

1 oz_f is a bit short of 0.3 N
1 lb_f is a bit more than 4 N
1 kg_f is a bit short of 10 N

Energy and Torque

Energy

Energy as work is the product of force times distance; thus its SI unit is N·m. The unit was given the name joule (after the British scientist) and the symbol J.

The joule is a unit of any kind of energy, be it work, heat, nuclear energy, kinetic energy, etc. Thus the joule replaces all energy units, be it ft-lb, lb-in, Btu, cal, Cal, kWh, eV, Q, or erg; in all, over fifty units.

Having an identical unit for any kind of energy introduces welcome simplification to calculations involving heat, electricity, etc.; no conversion (equivalency) factors are needed. Calculation of energy from power is also simplified and therefore easier to understand (see the next section). In nutrition, the use of joule eliminates the confusion that existed with the nutritional calorie being actually a kilocalorie.

The following figures show the amount of work produced by 1 J, the equivalence between work and heat of 1 kJ, and examples of the modern food-energy labeling.

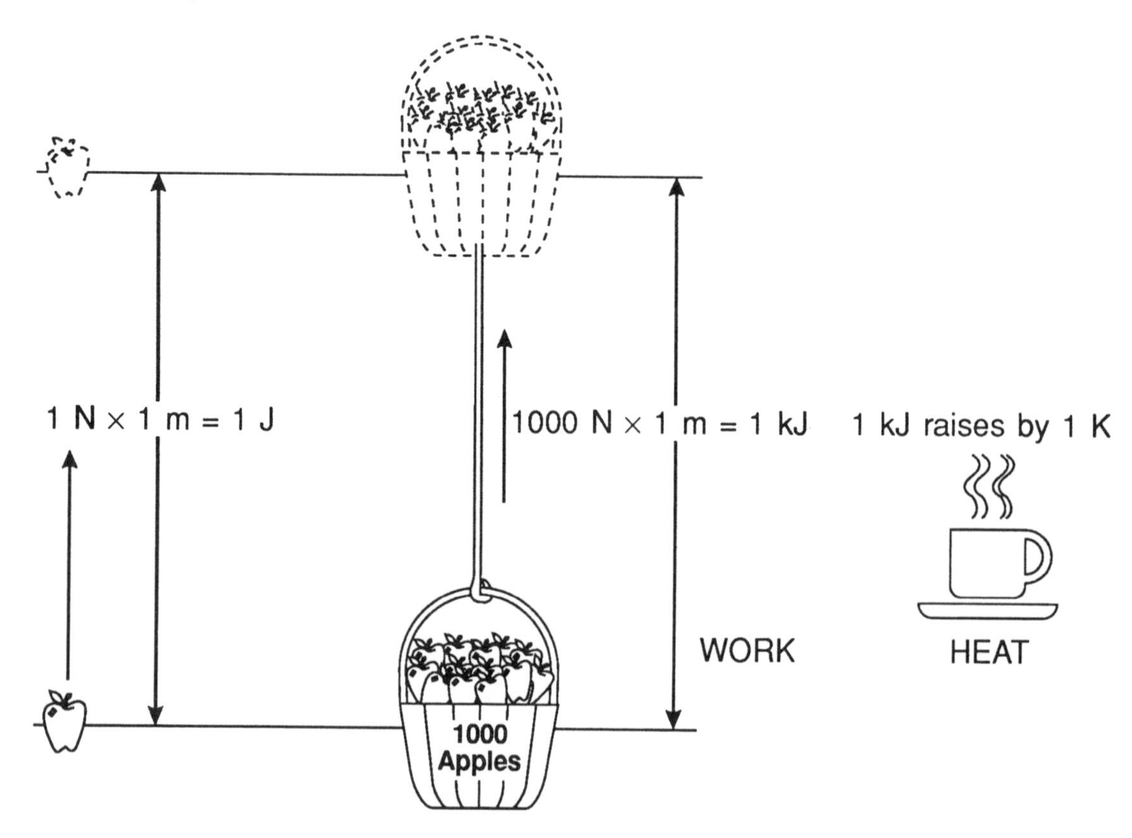

HOMOGENISED PASTEURISED SKIM MILK

SOUTHERN FARMERS COOP. LTD. F12
212 PIRIE ST. ADELAIDE, SOUTH AUSTRALIA, 5000.
REG. No. 813 TELEPHONE (06) 354 9600

PRODUCT OF AUSTRALIA

STORE BELOW 4°C

NUTRITION INFORMATION
SERVING SIZE 250ml/ 4 PER PACK

	PER SERVING	PER 100 ml
Energy	425 kJ	170 kJ
Protein	10 g	4.0 g
Fat	0.25 g	0.1 g
Carbohydrate	15 g	6 g
Sodium	148 mg	59 mg
Potassium	475 mg	190 mg
Calcium	400 mg	160 mg

Torque

Torque is also a product of force and distance. It, however, should not be expressed in joules. Physically, there is a difference between energy and torque, and that is reflected in retaining the symbol N·m for torque.

Strictly speaking, there is also a physical difference between torque in motion and torque in torsion or bending. This difference could be recognized by assigning N·m/rad to the "rotating" torque and N·m to the "static" torque. This is not done for reasons too detailed to go into here (a partial explanation is given in the section on Angle). It suffices to say that the symbol N·m/rad is considered the unit of torsional elasticity, not torque.

A Feel For Sizes

1 J: small apple lifted 1 m (potential energy)
1 kJ: content of a milk droplet (food energy)
1 MJ: slice of a pumpkin pie (food energy)

Equivalent Common Values

Energy:	1 Btu is almost exactly 1 kJ
	1 cal is a bit over 4 J
	1 Cal is a bit over 4 kJ
	1 kWh is exactly 3.6 MJ
Torque:	1 lb_f-ft is about 1.4 N·m

Power

Power is defined as the time rate of energy consumption or generation. Therefore, its SI unit is J/s. The unit was given the name watt in honor of the Scottish scientist who, ironically, coined the unit horsepower. The symbol of the unit is W.

The watt replaces not only the horsepower, but also all other power units such as Btu/h, cal/sec, ft-lb/min, metric hp, Btu/min, etc.

The figure below demonstrates the simplifications possible with SI in calculating the power of a burner.

Furnace Output

Values Required for Calculation:

SI		U.S.
44.2 kJ/g	Heating value of fuel	19,000 Btu/lb
0.89 g/s	Furnace fuel flow	7.05 lb/h
—	Conversion of time	60 min/h
—	Conversion of heat to work	778 ft-lb/Btu
—	Conversion of work rate to power	33,000 ft-lb/min-hp

Calculation:

SI

P = energy/time = 44.2 kJ/g × 0.89 g/s = 39.3 kW

U.S.

P = energy/time = 19,000 Btu/lb × 7.05 lb/h × 1/60 h/min
× 778 ft-lb/Btu × 1/33,000 min-hp/ft-lb = 52.6 hp (39.3 kW)

The conversion among various power units is complicated by the fact that there are differences in the size of some of them. For example, with horsepower there is the UK horsepower, metric horsepower, electrical horsepower and others, all differing slightly.

A Feel For Sizes

1 W: flashlight light bulb
1 kW: a two-horse team; small heater
1 MW: Indy race car engine

Equivalent Common Values

1 hp is about 0.75 kW
1 Btu/min is a bit under 18 W
1 ft-lb_f/sec is just over 1.3 W
1 cal/sec is a bit over 4 W

Remark: A Review of Newton, Joule, and Watt Derivation

Force:	N	[kg·m/s^2]
Energy:	N·m = J	[kg·m^2/s^2]
Power:	J/s = W	[kg·m^2/s^3]

As a summary of the terms and units that led to the derivation of the watt, the figure below demonstrates the relationship among gravity force, torque, work in rotational motion, and power in a mechanical and electrical device.

Gravity Force (mass × gravity acceleration)

The gravity force of a 102 kg man is 1 kN on Earth, since 102 kg × 9.8 m/s^2 = 1000 N, or 1 kN.

Torque (force × distance)

The torque the 1 kN force exerts if applied to the end of a 1 m long wrench is 1 kN·m, since 1 kN × 1 m = 1 kN·m.

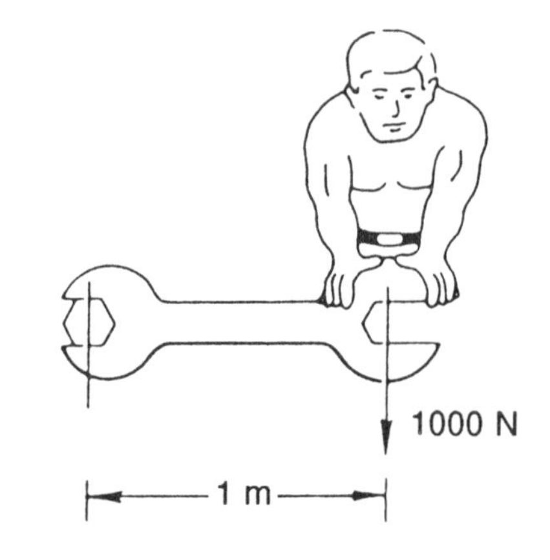

Work (force × distance)

The work the man would produce by pushing with the 1 kN force while walking once around along a circle of the 1 m radius is 6.28 kJ, since 1 kN × 2π × 1 m = 6.28 kJ.

Power (time rate of work)

The power the man would be generating "walking" around at a rate of 35 times a second is 220 kW, since 6.28 kJ × 35 s^{-1} = 220 kW.

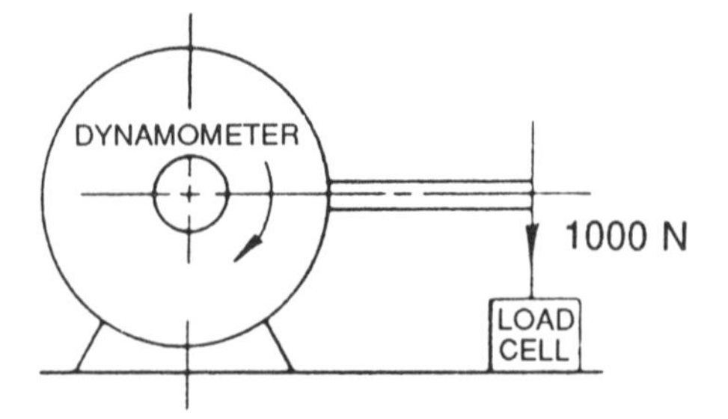

This power could be observed with a DC dynamometer on its voltmeter and ammeter. Multiplying the voltage and current readings would give 220 000 V·A, which is 220 kW.

Pressure

Pressure is defined as the action of a force spread over an area. Therefore, its SI unit is N/m2. The unit was given the name pascal in honor of the French scientist. The symbol of the unit is Pa.

The pascal is also the unit of stress, vacuum, etc. It replaces all other pressure units, including the unit called bar which was introduced only a few years before pascal was incorporated into SI.

Pressure ranges widely among the various disciplines of engineering and science, and prefixes therefore almost always accompany the unit. For example, whereas millipascals are convenient in acoustics and vacuum technology, kilopascals are convenient in weather data, and gigapascals in material sciences.

The figure below shows the derivation of the formula and unit for the pressure at the bottom of a liquid column, and an example of manometers graduated in kPa.

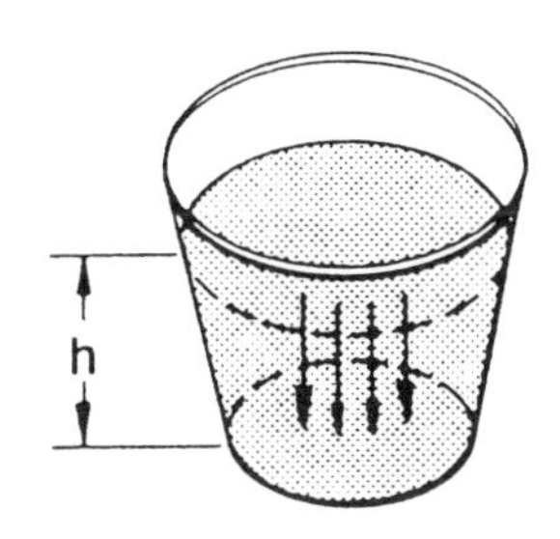

COLUMN PRESSURE

p = gravity force/area
= h × density × grav. accel.
(Pa) = (m × kg/m³ × m/s²) = (N/m²)

$$p = \text{gravity force/area} = h \times \text{density} \times \text{grav. accel.}$$

$$(\text{Pa}) = (\text{m} \times \text{kg/m}^3 \times \text{m/s}^2) = (\text{N/m}^2)$$

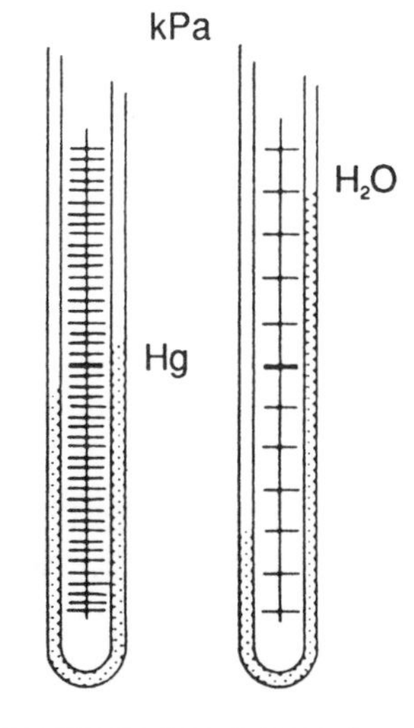

MANOMETER SCALES
FOR MERCURY AND WATER

In SI, the distinction between absolute pressure, gage pressure, and vacuum must be stated in words (there are no symbols for this). Examples:

- absolute pressure of 170 kPa
- gage pressure of 90 kPa; 90 kPa above ambient
- 90 kPa below ambient (the term vacuum is best avoided)

The pressure unit is also one of the units where the difference between a scale and an increment may cause an error in conversion; it is often not apparent what was meant by the term "atmosphere," for example. The scale conversion is complicated by the existence of two standardized reference air pressures, the “international” (101.325 kPa) and the “technical” (98.067 kPa).

The reference ambient (atmospheric, barometric) pressure level is expected to be set, for engineering purposes, at 100 kPa; the increment is, of course, the pascal.

In most engineering materials, stress is a two- or three-digit number if expressed in MPa. MPa is numerically equal to N/mm^2, which is handy with metric drawings where parts are always dimensioned in mm. Older literature often displays the N/mm^2 symbol in place of the preferred MPa.

A Feel For Sizes

1 Pa: loud sound; 0.1 mm of water
1 kPa: drop across a filter; 100 m in elevation
1 MPa: ten times the atmosphere
1 GPa: stress in a high-strength bolt

Equivalent Common Values

1 psi is a bit less than 7 kPa
1 kp/cm^2 is about 100 kPa
1 st. atm. is exactly 100 kPa
1 bar is exactly 100 kPa
1 inHg is about 3.5 kPa
1 mmH_2O is about 10 Pa

Frequency

There are three kinds of frequencies, and correspondingly three different units:

— Angular Frequency, commonly called angular velocity. Its unit is rad/s.

— Cycle Frequency. It is defined as the number of periodic events (cycles) per second. This unit was given the name hertz (in honor of the German scientist), symbol Hz.

— Rotational Frequency, commonly called speed of rotation or simply speed. Its unit is 1/s (s^{-1}).

The units of the cycle frequency and rotational frequency are sometimes, and incorrectly, written as c/s or cps, and r/s or rps, respectively. These symbols could be misunderstood in non-English speaking countries. Furthermore, cycles and revolutions are not units; if used, these words should be spelled out, or clearly abbreviated such as rev./s.

A Feel For Sizes

60 Hz: line frequency
530 to 1600 kHz: AM radio
88 to 108 MHz: FM radio

Equivalent Common Values

1 cps is 1 Hz
1 rev./sec. is 1 s^{-1}
1 rev./sec. is 2π rad/s

Acceleration

Acceleration is the time rate of velocity change. Thus the SI unit of longitudinal acceleration is m/s^2; the unit of angular acceleration is rad/s^2.

The unit m/s^2 replaces not only the in-lb equivalents, but also the ubiquitous "g."

Note: This "g," or "g force," was used in the past to signify a value of gravity acceleration (i.e., the acceleration of free fall; this is the downward acceleration all objects assume when dropped). The symbol g meant the physical quantity. The symbol g is best replaced by the symbol a_g, and the symbol g by its numerical value in m/s^2.

Here are the commonly encountered values of gravity acceleration on Earth:

— For rough estimates, $a_g = 10$ m/s^2.

— In standards work, and for reference, $a_g = 9.806\ 65$ m/s^2. (This is sometimes referred to as g_n.)

— In precision work, a_g is looked up from tables for the particular locality.

— For engineering purposes, in the U.S., $a_g = 9.80$ m/s^2, and in Europe, $a_g = 9.81$ m/s^2. The difference reflects on the more southerly location of the U.S.

There is more about gravity acceleration in the section on Force; the relationship between the kg_f and N is also explained there.

Equivalent Common Values

1 ft/sec^2 is about 0.3 m/s^2
1 "g" is almost 10 m/s^2

Vibration

The units of amplitude in longitudinal vibration are m for displacement, m/s for velocity, and m/s^2 for acceleration.

In the field of rotating machinery balancing, for example, one- or two-digit numbers result from using µm for displacement, mm/s for velocity, and m/s^2 for acceleration.

The figure on the next page displays the values of frequency, displacement, velocity and acceleration for some common machinery.

The method of conversion from the "vibration" in "g" to m/s^2 was presented in the section on Acceleration.

Equivalent Common Values

1 mil is about 25 µm
1 in/sec is about 25 mm/s
1 ft/sec^2 is about 0.3 m/s^2
1 "g" is about 10 m/s^2

TYPICAL AMPLITUDES OF VIBRATION IN SEVERAL APPLICATIONS

Courtesy of B&K
(including permission to modify as seen)

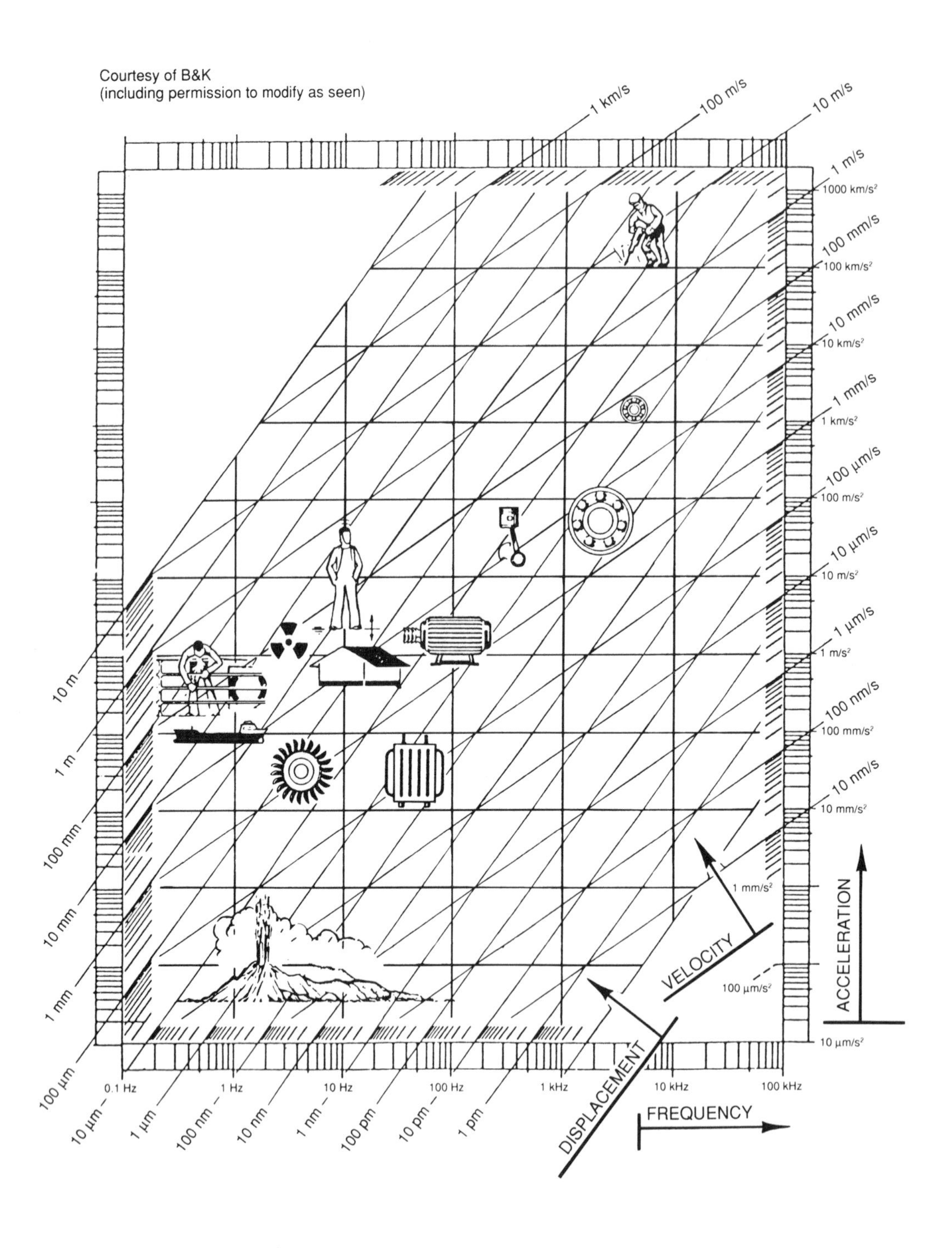

Other Quantities Used in Longitudinal Mechanics (alphabetical)

Elasticity (Stiffness)

Elasticity (stiffness, spring rate, spring constant) is defined as the increment of elastic deformation caused by a certain force. Accordingly, its SI unit is N/m. Conversion is a matter of realizing that the lb, oz, kg, etc., in the old systems are force units here, and thus the conversion factor for force must be used.

Impulse

Impulse is defined as the product of force and time. Accordingly, its SI unit is N·s. Conversion is a matter of realizing that the lb, oz, kg, etc., in the old systems are force units here, and thus the conversion factor for force must be used.

Momentum

Momentum is defined as the product of mass and velocity. Accordingly, its SI unit is kg·m/s. Conversion is a matter of applying mass (not weight!) and velocity conversion factors.

Note on Momentum and Impulse: In SI base units, momentum and impulse have the same unit. It is a reflection on the close relationship between the two quantities: The change in momentum is equal to the impulse.

Velocity

Velocity (speed) is defined as distance per time. Accordingly, its SI unit is m/s.

Density and the SI Terms for "Specific Weight" and "Specific Gravity"

Density is the ratio of one quantity to a unit of volume, area, length, or mass. There is power density (W/kg, W/m^3, ..), energy density (J/kg, J/m^3, ..), longitudinal density (kg/m), and a host of other densities whose use and units are a straightforward matter of substituting SI units for the non-SI ones.

Only *mass density* and *gravity force density* may present a problem because of the prior use of the terms "specific weight" and "specific gravity." A lot about this subject was covered in the section on Force.

"Specific Weight"

The ambiguity associated with the term "weight" as used by the public implies that "specific weight" may mean either:

— a certain amount of mass in a unit volume. This is specific mass, or *mass density* (often simply density) in SI.

— the gravity force of a certain amount of mass in a unit volume. This is specific gravity force, or *gravity force density* in SI.

The terms *mass density* and *gravity force density* clarify and replace the term "specific weight" in a manner similar to the way the terms *mass* and *gravity force* distinguish the two meanings of "weight." Thus, whereas g (gram) indicates mass, g/m^3 indicates mass density; whereas N (newton) indicates force (gravity force in this case), N/m^3 indicates gravity force density.

The value of either density depends on pressure, temperature, crystalline structure, etc., of the mass, while gravity force density depends also on the local gravity acceleration. Considering water, its mass density is 1 Mg/m^3, and its gravity force density is, say at the Washington DC latitude and altitude, 9.80 kN/m^3.

"Specific Gravity"

The term "specific gravity," which is also obsolete and has nothing to do with gravity, can best be viewed in the context of the double meaning of "weight." Thus "specific gravity" may mean either:

— the ratio of mass densities. This ratio is called the *relative mass density* in SI.

— the ratio of gravity force densities. This ratio is not used (it yields practically identical numbers as relative density).

Relative (mass) density is nondimensional. It is most often understood relative to water, in which case the relative density numbers are identical to the mass density numbers in Mg/m^3. (This may not be totally accurate in every case, but it is sufficiently accurate for all engineering purposes.)

Note: The ambient conditions for the determination of standard densities are nowadays often set at 100 kPa and 300 K.

A Feel For Sizes

Density: 1 kg/m^3: hot air
1 Mg/m^3: standard water
Relative density: 1: standard water
7.85: standard steel

Equivalent Common Values

1 lb_m/ft^3 is about 16 kg/m^3

Mass Density, Relative (Mass) Density, and Gravity Force Density

Mass	Mass Density kg/m³	Relative Mass Density -	Gravity Force Density*** N/m³
osmium	22 600	22.6 *	221 000
platinum	21 400	21.4 *	210 000
gold	19 300	19.3 *	189 000
mercury	13 600	13.6 *	139 000
lead	11 300	11.3 *	111 000
copper	8 950	8.95 *	87 800
steel	7 850	7.85 *	76 900
aluminum	2 700	2.70 *	26 500
glass	2 600	2.60 *	25 500
concrete	2 300	2.30 *	22 500
sulfuric acid	1 800	1.80 *	17 600
magnesium	1 770	1.77 *	17 300
battery acid	1 260	1.26 *	12 350
water	1 000	1 *	9 800
oil	900	0.90 *	8 820
alcohol	790	0.79 *	7 750
hardwood	700	0.70 *	6 860
softwood	500	0.50 *	4 900
styrofoam	20	0.02 *	196
butane	2.41	2.00 **	23.6
propane	1.83	1.52 **	17.9
oxygen	1.33	1.10 **	13.0
air	1.206	1 **	11.8
nitrogen	1.165	0.966**	11.4
helium	0.167	0.138**	1.64
hydrogen	0.084	0.069**	0.82

* relative to water
** relative to air
*** at the latitude and altitude of Washington DC ($a_g = 9.80$ m/s²)

Note: All numbers are rounded, and they represent the average values listed in several handbooks.

Viscosity

In SI, dynamic (absolute) viscosity has the unit Pa·s, and kinematic viscosity has the unit m^2/s, as it is dynamic viscosity divided by mass density.

Conversion of the now obsolete poise and stokes to the SI units is a matter of shifting the decimal point, since both were defined in terms of metric measurements. The Equivalent Common Values table below shows the relationship of those two units to the SI ones.

All the other units of viscosity, such as Redwood, Seyboldt Universal and Fural seconds are not SI; the conversion from these units is best done using tables of comparison.

Notice that dynamic and kinematic viscosities are numerically equal for standard water if expressed in mPa·s and mm^2/s, respectively.

Equivalent Common Values

1 centipoise = 1 mPa·s
1 centistokes = 1 mm^2/s

Moment of Inertia and Other Moments

There are several quantities referred to as moments in engineering. This section provides a brief survey of them.

The *moment of force* (couple) was discussed in the section on Energy and Torque.

The *moments of area* were mentioned in the section on the unit of length. To change from a non-SI unit to SI is a straightforward substitution of the SI units, while milli is usually the preferred prefix.

A similar, straightforward treatment is possible with the *moment of mass*.

Only the *moment of inertia* may present a problem and is discussed here.

Moment of Inertia

Moment of inertia (mass moment of inertia, angular inertia) is defined as mass times the radius squared ($I = m \times r^2$). Its SI unit is therefore $kg \cdot m^2$.

A variety of non-SI units is used for moment of inertia throughout the world, and the term itself is often used for the quantity properly called the second moment of area (unit m^4). Furthermore, the so-called Flywheel Effect is sometimes applied in place of the polar mass moment of inertia, and its unit differs among nations and companies. Before conversions are made, one must determine if the kg, g, lb, or oz in the unit refers to mass or to force. Without this determination and a use of proper terminology, conversions are easily misapplied.

Note: This is one of the many cases where the use of SI units removes a potential source of errors and the related confusion. With the moment of inertia, in particular, errors with non-SI units and with conversions are frequent.

When it is too difficult to find the necessary information that would indicate which conversion factor is proper, measuring the moment is sometimes the easiest way out. The following figure shows how to calculate the polar mass moment of inertia from experimental data and how to prepare the experiment. Note that a dimensional analysis does not come out "right" in the last equation. The equation has been simplified and, as a result, some units are "hidden."

Experimental Method for Determination of Polar Mass Moment of Inertia

Several methods have been derived for the experimental determination of polar mass moment of inertia. The one described here is accurate and easy to apply. All you need is a scale, measuring tape, lengths of wire, stopwatch, and a place to suspend the component as shown in the picture.

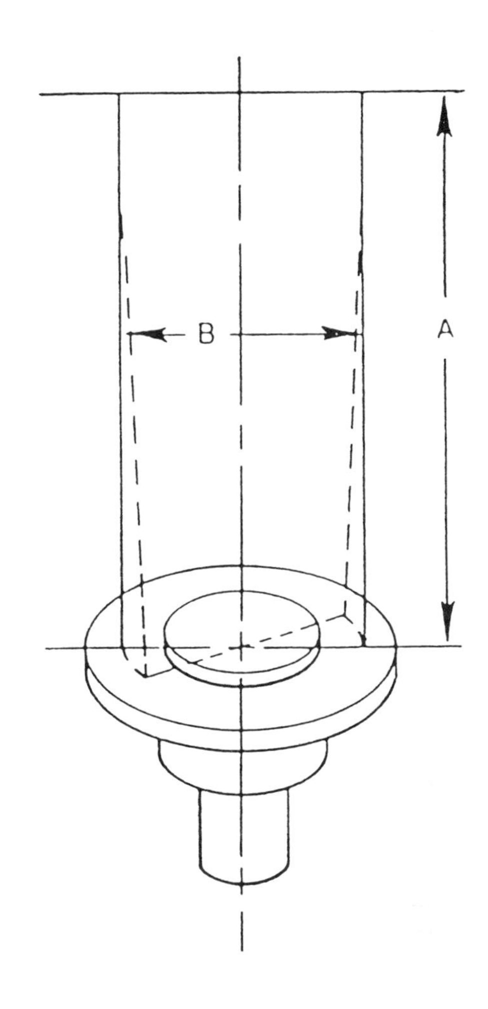

The distances *A* and *B* should be selected such that *A* (the length of the wires) is at least five times longer than *B* (the distance between the wires).

The assembly is rotated slightly around its axis and released. The cycle time (*t* in s) is determined.

Knowing the mass (*m* in kg), the lengths (*A* and *B* in m), we can calculate the moment (*I* in kg·m²) from the formula:

$$I = \frac{m \times a_g \times B^2 \times t^2}{16 \times \pi^2 \times A}$$

Realizing that the value of $a_g \approx \pi^2$ on Earth, the equation can be simplified to:

$$I \approx \frac{m \times B^2 \times t^2}{16 \times A}$$

Note: Simplified equations such as this one present a potential for error when they are used within systems of units other than the one in which they were simplified. To avoid the error, it is important to perform a dimensional analysis with all such equations before converting them.

Equivalent Common Values

1 lb_f-in-s^2 is about 0.1 kg·m^2
1 kg_f-cm-s^2 is about 0.1 kg·m^2
1 GD^2 [kg-m^2] is about 0.25 kg·m^2
1 WR^2 [lb-ft^2] is about 0.04 kg·m^2

Remark: A Survey of Quantities and Units in Rotation

Angle, Angular Velocity (Angular Frequency), and **Angular Acceleration**. The symbols of their units are rad, rad/s, and rad/s^2, respectively.

Angular Inertia (Mass Moment of Inertia, Rotational Inertia) is a property of the radial mass distribution of a body about an axis (mr^2). The symbol of its unit is kg·m^2.

Angular Force (Moment of Force, Torque, Bending Couple) is force times moment arm (lever arm) length. The symbol of its unit is N·m.

Angular Elasticity (Angular Stiffness, Torsional Stiffness, Torsional Spring Rate) is torque divided by angle of twist. The symbol of its unit is N·m/rad.

Angular Work (Rotational Work) is torque times angle of rotation. The symbol of its unit is J.

Angular Kinetic Energy (Rotational Kinetic Energy, Kinetic Energy of Rotating Body) is proportional to moment of inertia times angular velocity squared. The symbol of its unit is J.

Angular Momentum (Moment of Momentum) is longitudinal momentum (mv) times moment arm length, or moment of inertia times angular velocity. The symbol of its unit is kg·m^2/s.

Angular Impulse is the product of torque and the time it acts. The symbol of its unit is N·m·s.

Note: For the use of the symbol rad in the above units, see the section on Angle.

Amount of Substance

The term "amount of substance" is the SI expression for what used to be called gram molecule, gram mole, and the like.

The unit of the amount of substance is mole, symbol mol. Mole is defined as the amount of a substance of a system that contains as many elementary entities as there are atoms in 12 grams of the carbon isotope called carbon 12.

The elementary entities can be molecules, atoms, ions, electrons, and other particles or a group of particles. To avoid an error, the particles must be specified. The statement "the mass of 1 mol of water is 18 g" should be accompanied by a reference to molecules.

The distinction between amount of substance and mass is absolute; there is no "conversion factor" between them. Mass is a direct measure of inertia, the amount of substance is not. For a given magnitude of mass, the magnitude of the amount of substance can vary enormously. In referring to molecules, for example, and rounding off the numbers, 1 g of hydrogen has 0.5 mol, while 1 g of water has 0.055 mol. The amount of substance in 1 g of water, as another example, is ten times the amount of substance in 1 g of glucose.

There is no way to obtain a physical feel for the size of the mole, the feel so important for the ease of dealing in grams and newtons. The amount of substance cannot be measured directly as yet, as there is no suitable instrument available today that would count molecules, atoms, etc., directly. The values are calculated, and for many substances the values are not known.

SI brought a reform to the terminology of this physical quantity somewhat similar to the reform it brought to the terminology of the previously discussed mass and force. Where mass and gravity force terms distinguish and replace the two meanings in weight, amount of substance and its unit replace the older terms such as gram-ion, gram-molecule, and gram-atomic weight.

In this context it should also be mentioned that the word "weight" is replaced by the word "mass" when referring to molecular weight, atomic weight, and the similar pre-SI expressions, while, as before, the values are relative.

Electricity

Units of electricity and magnetism were metric from inception. Over the decades, some units were added, some were weeded out, and the definitions of some were modified, but most of these changes were made before SI was born. Still, units not incorporated into SI (e.g., maxwell, oersted) are seen sometimes, and therefore a review is provided here.

The review lists the obsolete units, their equivalent SI units, and the conversion factors.

Non-SI Unit	SI	Conv. Factor
abampere	A (ampere)	1.000×10^{1}
abcoulomb	C (coulomb)	1.000×10^{1}
abfarad	F (farad)	1.000×10^{9}
abhenry	H (henry)	1.000×10^{-9}
abmho	S (siemens)	1.000×10^{9}
abohm	Ω (ohm)	1.000×10^{-9}
abvolt	V (volt)	1.000×10^{-8}
ampere hour	C	3.600×10^{3}
faraday (chemical)	C	9.650×10^{4}
faraday (physical)	C	9.652×10^{4}
gauss	T (tesla)	1.000×10^{-4}
gilbert	A	7.958×10^{-1}
maxwell	Wb (weber)	1.000×10^{-8}
mho	S	1.000×10^{0}
oersted	A/m	7.958×10^{1}
ohm circ. mill per foot	$\Omega \cdot mm^2/m$	1.662×10^{-3}
statampere	A	3.336×10^{-10}
statcoulomb	C	3.336×10^{-10}
statfarad	F	1.113×10^{-12}
stathenry	H	8.988×10^{11}
statmho	S	1.113×10^{-12}
statohm	Ω	8.988×10^{11}
statvolt	V	2.998×10^{2}
unit pole	Wb	1.257×10^{-7}

To convert from the non-SI units to SI, multiply the non-SI value by the conversion factor.

Note: The units with names preceded by EMU (electromagnetic unit) or ESU (electrostatic unit) are now also obsolete.

Light

The changes that SI brought to the measuring units in optical sciences were few. The units lambert, footcandle, and several others were weeded out, and some definitions changed, but not all this happened as a result of the establishment of SI.

Several years after SI was introduced, the definition of the candela was modified to be based on monochromatic radiation of a certain frequency and power density.

Here is a brief survey of the common light quantities and their units in SI:

Quantity	Unit Symbol (Name)
luminous intensity	cd (candela)
luminous flux	cd·sr = lm (lumen)
illuminance	cd·sr/m^2 = lm/m^2 = lx (lux)
luminance	cd/m^2

Common light bulbs are rated by their energy consumption rate (power) and the quantity of light produced (luminous flux). As a ballpark figure, a 75 W light bulb emits about 1 klm.

Radiology

Radiology-related quantities and units are listed in the table below. The table also shows the now-obsolete units roentgen, rad, rem and curie, and it demonstrates in the column on the right how the SI units with special names relate to other SI units.

For details of the definitions and units in radiology, the reader is referred to, particularly, the International Commission on Radiation Units (ICRU) publications.

Quantity	Unit Symbol (Name) Obsolete	SI	Expressed in other SI units
activity	Ci (curie)	Bq (becquerel)	s^{-1}
exposure	R (roentgen)	——	$C{\cdot}kg^{-1}$
absorbed dose	rad (rad)	Gy (gray)	$J{\cdot}kg^{-1}$
dose equivalent	rem (rem)	Sv (sievert)	$J{\cdot}kg^{-1}$

For reference:

$1\ Bq \equiv 27\ pCi$
$1\ C/kg \equiv 3876\ R$
$1\ Gy \equiv 100\ rad$
$1\ Sv \equiv 100\ rem$

Chapter 4

Non-SI Metric Units and Terms

Metric units and terms that are not SI or are outright obsolete are still frequently used today. The following is a list of the more common ones, arranged in alphabetical order. Most were mentioned in the preceding pages.

- are, symbol a or dam^2 — This unit, also spelled ar, was a unit of area equalling 100 m^2. The unit is obsolete, surviving only in the prefixed form as the hectare (see below). Note that the symbol a means year nowadays.

- cc — See millilitre (mL, ml).

- g — This is a symbol for the unit gram. Previously, g was also a unit of acceleration, which in SI is the m/s^2. Observe that *g* (in *italics*) denotes the *quantity*, not the unit. To avoid confusion, *g* is best replaced by a_g.

- hectare, symbol ha or hm^2 — This is a unit of area, commonly used in many metric countries. One hectare is 100 are, as the prefix hecto implies. The unit should be avoided; instead, m^2 or km^2 should be used: 1 ha = 10 000 m^2, 100 ha = 1 km^2.

- kilo — As a word by itself, kilo is often used in common parlance for the word kilogram. Kilo is the prefix, of course, for 1000 of any unit in SI.

- litre (liter, litr, λυτρ, etc., symbol L, l, ℓ) — This is a special name and symbol for dm^3. Previously, there was a difference, ever so slight, between the volumes of L (l, ℓ) and dm^3; today the distinction no longer exists. Regardless, only the dm^3 symbol should be used in engineering as it is difficult to deal with L (l, ℓ) in dimensional analysis, because the relationship of L, mL to dm^3, cm^3 is often confused even among engineers, and because the "el" symbols can be misinterpreted, particularly the lower case l.

- micron — This is an obsolete name. Originally meaning 1 μm, it is sometimes used to mean a millionth of an inch. It also has other meanings. The term is not used in SI.

- millilitre (milliliter, etc., symbol mL, ml, mℓ) — This is a special name and symbol for cm^3, popularly known in this country as cc. Both ml and cc should be avoided; beakers, for instance, should display the cm^3 symbol.

- ton (tone, tuna, tonne) — As a name of a unit, the ton has at least eight different meanings. These are among them:

  • metric ton: - as a mass unit it means Mg
  - as a force unit it means kN

  (An example: a 1 Mg object exerts approximately 10 kN gravity force on Earth.)

  • ton in refrigeration is a unit of power (W). (One such ton is about 3.5 kW.)

  • ton in shipping is a unit of volume (m^3). (One such ton is about 2.8 m^3.)

  There are also long ton, short ton, etc. It is apparent that unless you are familiar with the particular discipline, the use of the term can lead to errors. Ton should be replaced with the appropriate SI term and symbol in all cases.

Chapter 5

Rules for Writing Units and Numbers in SI

The following rules were established to facilitate data communication. They are similar in all countries and in all languages. Units and numbers presented according to these rules need neither translation nor interpretation. The uniformity saves time and helps to prevent mistakes.

The rules represent a compromise intended to suit all languages, to ease arithmetic manipulations, to prevent ambiguity, and to retain some of the traditions of the metric system. Once learned they are easy and natural to follow. They were adhered to in this book faithfully and most likely helped your understanding of the text.

Symbols of units and prefixes are always written singular (5 g not 5 gs), as symbols do not vary by definition. Symbols are also not followed by a period unless grammar dictates otherwise (end of a sentence). Symbols are international: They are written the "SI way" no matter what script a country uses for its native language(s), and they are always written upright (as distinguished from *italic*).

Beware that some letters symbolize different meanings when written capitalized or when in lowercase (G-giga, g-gram). Uppercase letters are used for symbols of prefixes that represent multiples larger than kilo (M, G, ...), and for symbols of units derived from a proper name (N-newton, Ω-ohm, Hz-hertz, Pa-pascal; note that only the first letter is uppercase in these two-letter symbols).

The prefix setting is with no space between it and the unit. This applies to both the symbol and name. Example: mg and milligram are correct; m g and milli gram or milli-gram are not.

Names of units may be written either singular or plural, and they are always written in lowercase: joule is a name of a unit, while Joule is the name of the scientist; similarly hertz and Hertz. Notice that the unit would have to start uppercase if beginning a sentence; it is best to write sentences in a way to avoid such placement. Also, you should avoid writing out the names of units and use symbols wherever possible; such practice saves time and space, eliminates the possibility of spelling errors, and facilitates understanding of data in all languages.

Multiplication (among unit symbols) is indicated by inserting a raised dot or by leaving a space between the symbols (N·m or N m, but not N.m, N-m, N×m, etc.). In this country, the raised dot notation is more common. Compare these notations with the above-mentioned "prefix setting," and notice that when the space notation is used, sloppy writing could lead to an error; the symbol for mm, for instance, written with a space means m^2. (Note: The on-line dot also indicated multiplication in the past; it was ruled out for this function in 1992.)

Division (between unit symbols) is indicated by a solidus (/), a horizontal line (-), or a negative exponent ($^{-1}$). There is only one solidus in a unit, regardless of how many symbols there are in the numerator or the denominator (W/m·K, not W/m/K). To avoid all ambiguity, units in the denominator should be placed into parentheses. Examples: W/(m·K), W/(m^2·sr).

The unit in the denominator (of a derived unit) should not contain a prefix. The number of digits should be adjusted by placing a prefix with the unit in the numerator, not denominator. For instance, the elasticity of a spring may be expressed as 5000 N/m, or 5 kN/m, but not 5 N/mm. Similarly, the velocity of 2 km/s is preferred over the equivalent 2 m/ms. However, exceptions to this rule are often made to accommodate convention, convenience, the significant number of digits rule, and kg (kg is a base unit).

A number followed by a unit is written with a space between the number and unit (35 mm). In this country, a hyphen instead of the space is often used when the number serves as an adjective (35-mm film). That usage is not SI.

Long numbers are grouped into three-digit sections separated by a space, or a half-space (86 400; 0.002 35). The comma is not to be used for this function except in monetary transactions. A four-digit number need not have the space (4000); it is sufficiently short to be read easily. If it were a part of a column of numbers, it would, of course, have the space. This spacing convention eliminates the possibility for a mistake in those parts of the world where a comma is a decimal marker. Notice that the three-digits spacing matches the preferred 1000 factor in the prefix system.

The decimal marker can be either a comma or an on-line dot. In this country the on-line dot is preferred. Compare these notations with the multiplication symbols and with the spacing in long numbers. (Note: It is expected that the comma only will be approved with the next revision of the rules.)

The decimal zero: A zero is placed in front of the decimal marker when a number is smaller than one. This is done regardless of whether the marker is a comma or a dot. Example: 0.3 kg or 0,3 kg are both acceptable at this time, but .3 kg or any other is not.

Typing SI Symbols

Here are tips for typing the "metric" characters that may not be available on some typewriters or keyboards. The characters are the exponents ^{1}, ^{2}, ^{3}, $^-$, the raised dot ·, the symbol for degree °, and the Greek letters μ and Ω.

On typewriters:

Several typewriters have those characters as an option. Notably the IBM on Courier 10 (No. 1167238) and Letter Gothic 12 (No. 1167239).

If your typewriter cannot be so equipped, consider the following alternatives.

The exponents, raised dot, and degree symbol can be typed by rolling the platen a half notch. The character ° can also be spelled out (degree) or abbreviated (deg.) instead.

Instead of the Greek μ the letter u can be typed and the "leg" added to it by hand; Ω can be spelled out (ohm).

On word processors:

The exponents, raised dot, and degree symbol can be entered as superscripts. In addition, on word processors using the IBM extended graphic character set, ^{2} can be entered as Alt 253, · as Alt 250, ° as Alt 248.

In the case of the Greek characters, and using the IBM set, μ can be entered as Alt 230, Ω as Alt 234.

Note on ℓ: In some countries the letter ℓ represents dm^3; similarly mℓ represents cm^3. The trouble with typing this letter can be avoided by typing the SI symbols dm^3 and cm^3 instead.

Note on typing written-out names vs. symbols: Avoid writing out the names of units and prefixes. Type symbols instead (1 kg, not one kilogram; 1 mA, not one milliampere, etc.). Using symbols saves time, space, eliminates misspellings, and makes reading easier. Consider also that the spelling may differ among nations.

SI Symbols with "Uppercase Only" Printers

This is a list of symbols which substitute for the characters not available on "CAPS ONLY" printers.

Units:

Name	Proper Symbol	Substitute Symbol
metre, meter	m	M
gram	g	G
second	s	S
candela	cd	CD
mole	mol	MOL
radian	rad	RAD
steradian	sr	SR
hertz	Hz	HZ
pascal	Pa	PA
ohm	Ω	OHM
weber	Wb	WB
lumen	lm	LM
lux	lx	LX
becquerel	Bq	BQ
gray	Gy	GY
sievert	Sv	SV

Prefixes:

Name	Proper Symbol	Substitute Symbol
giga	G	G
mega	M	MA
kilo	k	K
deci	d	D
centi	c	C
milli	m	M
micro	µ	U
nano	n	N

Exponents:

s^2 can be substituted by S2

s^{-2} can be substituted by S-2

kg^2 can be substituted by KG2

Multiplication Dot:

N·m can be substituted by N M

$kg \cdot m^{-3}$ can be substituted by KG/M3

Chapter 6
Converting Units and Rounding Numbers

The rules for converting and rounding are identical in any system of units, and the same rules apply for converting among units of the same system as among units of differing systems. Within SI *no conversions are required. Within the in-lb system, on the other hand, conversions are taken for granted, for example, conversions between feet and inches, between horsepower and watt, between* inHg *and* psi, *between* ft-lb *and* Btu, *and so on.*

Since the in-lb system is still firmly ingrained in this country, converting and rounding should not need to be introduced as a part of the metric changeover. Converting between in³ *and* fl.oz. *is not, in principle, any different than converting between* in³ *and* cm³. *Despite that, this subject is often requested in training seminars, and thus a brief review is provided here accompanied by tables of conversion factors and reference numbers.*

The use of the conversion factors table (Table 2) is identical to the use of tables which list conversion factors for units of the in-lb system. To provide additional assistance, the third page of Table 2 shows in the bottom right-hand corner two examples of how to do the calculations.

The usual problem facing those who perform conversions is not so much in the application of the conversion factors, as there is nothing particularly metric about it, but rather in deciding which factor to use. The preceding chapters presented the material needed for acquiring the knowledge to select the right factor.

The usual problem facing those who read the converted numbers is, on the other hand, the lack of common sense often exhibited in the rounding of the conversion results. Although here, too, the subject is no more metric than inch-pound, somehow numbers converted to metric are often presented with insignificant digits. The further text is therefore devoted mainly to the rounding issue. It includes the description of a rule for rounding which predominates globally.

When a value is converted from one unit to another, its number of significant digits must reflect the accuracy implied in the original number. Accuracy must not be sacrificed, but also not exaggerated. To determine how many digits are significant, one must understand the circumstances involved in the determination of the original value and the "accuracy message" the number carried.

For example, a dimension of, say, 20 in. may mean a length of between 19 to 21 in. if it refers to a piece of firewood, or 20.000 in. if it refers to a part of a machine. Such a dimension is just as correctly converted to 1/2 m as it is to 508.00 mm. Without the knowledge of the history and purpose of the original number, you cannot decide which converted value is appropriate.

As another example, consider bolt torque. The relationship of the tightening torque to the clamping force, and the accuracy of the common torque wrenches implies some 40% variability. A 15-18 ft-lb torque is, therefore, correctly converted to 20-25 N·m. A more accurate value, say, 20.3-24.4 N·m, is unprofessional, as it contributes nothing to the bolted joint quality while making the data more difficult to read and remember.

Promotional literature necessitates even more liberal rounding to facilitate comprehension. For example, a big car engine power is best listed rounded to the nearest 10 kW in advertising brochures.

With fractions of an inch, the nearest 1 mm or at most 0.5 mm is close enough. For example, a pin protrusion 1/32-1/16 in. should become 1-1.5 mm.

In converting drawing tolerances, give the tolerance a hard look. Can it be loosened a bit? Always take this option if it is available to you. Remember that the original tolerance was most likely rounded off "inward," and is thus tighter than it needs to be. Converting and then rounding such tolerance further "inward" may increase the part's cost. For example, the new tolerance may require grinding where milling sufficed.

The rule to remember is that conversions must be rounded off to reflect the accuracy which was implied in the original value, thereby to deliver the message the number of significant digits was to convey. The choice of the simplest numbers and a consistent use of symbols should always be given preference as this contributes to ease of reading and retention.

Finally, beware of constants in equations when converting mathematical expressions. Many an equation has been simplified by lumping conversion factors into a constant and as a result some units are "hidden." Simplified equations present a potential for error when they are used within systems of units other than the one in which they were simplified. To avoid the error, it is important to perform a dimensional analysis before the equations are converted. Ideally, there should not be any constants in an equation to be converted except the "natural" ones, such as π.

Remarks Concerning the Rules of Rounding

The rules for rounding and the related mathematics is the same in the metric system as in any other system. This topic is not germane to metric, nor is the mathematics often needed in engineering conversions. On the contrary, its blind application may cause a loss of accuracy rather than a gain, the time spent notwithstanding.

Instead of the rigorous mathematical treatment, it is more important, and usually sufficient, that common sense supported by the in-depth knowledge of the origin and purpose of the to-be-converted numbers be applied as discussed earlier; the rest will fall into place naturally.

A lot of rounding can be avoided altogether. For example, in solving equations, one should arrive at the final form algebraically and only then enter the numbers. Entering them earlier may result in a loss of accuracy, both implied and real, as the rounding error is entered over and over again.

As for the number of significant digits in the metric measures, remember that prefixes make that rule easy to apply: compare 2000 kW versus 2 MW.

As to the rounding of the number 5, there is one possible deviation in the method of rounding. In the U.S., and a few other countries, a number ending with 5 may be rounded either (1) upward, or (2) upward or left unchanged depending on whether the digit preceding the 5 is odd or even. Worldwide, the latter method was abandoned. Examples: 1.35 is rounded to 1.4; identically, 1.65 is rounded to 1.7.

The following tables contain a listing of SI and other units, reference numbers and conversion factors.

Table 1 lists quantities, examples of usage, various units, and ballpark figures for typical mechanical engineering applications. Table 2 lists conversion factors. A detailed description is at the heading of each table. For conversion of units in electricity and radiology see the sections on Electricity and Radiology in Chapter 3.

Table 1. Quantities, Units and Typical Values

This is a table of old units and SI units arranged in the alphabetical order of their respective physical quantities. About fifty quantities are listed, selected to cover disciplines commonly used in mechanical engineering and daily life, particularly where the selection of a conversion factor may pose a problem.

Experience indicates that the resistance against using data and calculations in SI stems from the lack of the feel for the "ballpark figures." To provide some feel, the table includes a column of Typical Values with approximately one hundred and fifty constants and reference numbers.

The column of the SI units shows their symbols in the form best suited for typing and also in the prefix-free form. The entries in the Typical Values column provide a guide for the selection of a suitable prefix.

Quantity	Example of Usage	Example of Old Units	SI Units	Typical Values
Acceleration: angular	shaft, governor, centrifuge	deg/sec^2, rad/sec^2	rad/s^2	centrifuge bucket: 40 rad/s^2
longitudinal	gravity, vehicle, valve train	ft/sec^2, g, in/sec^2	m/s^2	Earth gravity: 9.8 m/s^2; dragster: 20 m/s^2; cam follower: 1000 m/s^2
Angle	geometric pictures, slope	deg, min, sec, %	rad [deg,%]	one radian equals 57.3 degrees; a one in two slope equals 50%
	twist, dynamics of rotation	rad, degree	rad	6.28 rad equals one revolution
Area	cross section, land measure	in^2, ft^2, acre	m^2	office space per person: 10 m^2; writing desk: 1 m^2; paper clip wire: 1 mm^2
Coeff. of therm. expansion	shrink fit, volumetric growth	1/°F	1/K	Al alloy: 22 μm/m·K; steel: 11 μm/m·K; kerosene: 1 dm^3/m^3·K; water: 0.18 dm^3/m^3·K
Conductivity	heat transfer, insulation	Btu-in/h-ft^2-°F	W/m·K	asbestos: 0.07 W/m·K; concrete: 1 W/m·K; steel: 62 W/m·K
Consumption: general	See Flow			
specific	brake spec. fuel consumption	lb/hp-h	g/J	diesel engine: 80 g/MJ; gasoline engine: 100 g/MJ
Density: gravity force	gravity force per unit volume	lb/ft^3	N/m^3	air: 12 N/m^3; oil: 8.5 kN/m^3; water: 9.8 kN/m^3 (on Earth)
mass	mass per unit volume	lb/ft^3, lb/gal	g/m^3	air: 1.3 kg/m^3; oil: 870 kg/m^3; water: 1000 kg/m^3; Al alloy: 2700 kg/m^3; steel: 7850 kg/m^3
power	area radiation, energy transfer	Btu/ft^2-h	W/m^2	solar radiation on Earth: 1.4 kW/m^2
relative	ratio of densities	non-dim.	non-dim.	relative mass density of water is 1, of steel 7.85, if water is the reference
Dimension	engineering design	in	m*	shaft dia.: 23.410; shaft dia. tolerance: ±0.004; car length: 5730. See footnote*.
Elasticity: angular (torsional)	spring torque per increm. of angle	deg/100 ft-lb	N·m/rad	soft driveshaft: 600 N·m/rad; rigid driveshaft: 300 kN·m/rad
longitudinal	spring force per increm. of length	lb/in	N/m	light spring: 1 kN/m; car suspension: 30 kN/m; railroad car spring: 4 MN/m
Energy: content	heat content in fuel, food	Btu/lb, Cal/g	J/g	heating oil: 42 MJ/kg; methanol: 20 MJ/kg; protein: 17 kJ/g; fat: 39 kJ/g
general	work, heat, electricity	ft-lb, Btu, kW-h	J	60 W bulb: 5 MJ in one day; diet soft drink: 4 kJ; 1 MJ of electricity costs 2¢
specific	heating, cooling	Btu/lb-°F	J/g·K	tin: 0.23 kJ/kg·K; iron: 0.45 kJ/kg·K; oil: 1 kJ/kg·K; water: 4.2 kJ/kg·K
Flow: mass	fuel flow rate	lb/min	g/s	car engine fuel flow: 2 g/s; rocket engine: 100 kg/s
volume	air flow rate	ft^3/min, yd^3/min	m^3/s	air usage of marathon runner: 1 dm^3/s, car engine: 100 dm^3/s
Force: general	drag, thrust, load, strength	oz, lb, ton	N	strength of an M10 low-grade steel bolt: 10 kN
gravity	weight, load	oz, lb, ton	N	102 kg mass exerts 165 N force of gravity on Moon, 1 kN on Earth, 275 kN on Sun
Frequency: angular	natural frequency	deg/sec, deg/min	rad/s	100 rev./s equals 628 rad/s
cycle	vibration, pulsing	c/min, c/sec	Hz	electricity: 60 Hz; middle "C": 261.6 Hz; an AM radio: 800 kHz, FM: 90 MHz
rotational	rpm, rps	r/min, r/sec	1/s	induction motor: 30 1/s; small generator set: 60 1/s; turbocharger: 2000 1/s
Heat	See Energy			
Length	distance; see also Dimension	mil, in, ft, yd	m	dime thickness: 1 mm; long pace: 1 m; NY to Detroit: 1000 km
Load	See Mass, or Force, or Power, etc.			
Mass	weighing, load, inertia	oz, lb, ton, slug	g	1 dm^3 contains 1 kg of water; average person: 70 kg; class 8 truck: 35 Mg
Modulus of: elasticity	longit. (E), shear (G), volumet. (K)	lb/in^2	Pa	steel: E = 200 GPa, G = 80 GPa; oil: K = 1.5 GPa
section	bending and twist calcul. (W)	in^3	m^3	rod of 10 mm dia.: W = 100 mm^3 in bending, 200 mm^3 in torsion

Moment of: area, 1st	centroid of shape	in^3	m^3	1 m × 1 m plate: 0.5 m^3
area, 2nd	bending and twist calculation	in^4	m^4	ϕ10 mm rod: 500 mm^4 in bending, 1000 mm^4 in torsion
force	torque, bending couple	ft-lb, in-lb	N·m	small car engine torque: 100 N·m; M8 nut tightening torque: 8 N·m
inertia	dynamics of rotat., Flywheel Effect	lb-in-sec^2, lb-ft^2, slug-ft^2	g·m^2	polar of a ϕ100 mm steel disk 10 mm thick: 0.75 g·m^2, ϕ1000 mm: 7.5 kg·m^2
mass	balance, vibration	lb-in, oz-in	g·m	small armature unbalance: 5 mg·m
Momentum: angular	rotation, Moment of Momentum	slug-ft^2/sec	g·m^2/s	ϕ1000 mm steel disk 10 mm thick at 10 1/s: approx. 500 kg·m^2/s
longitudinal	kinetics, motion	lb-in/min	g·m/s	1 kg object moving at 1 m/s: 1 kg·m/s
Power	heating, cooling, output, energy rate	Btu/min, hp	W	man: 100 W cont.; small kerosene heater: 2000 W; house furnace: 40 kW
Pressure	stress, vacuum, injection	psi, inHg, inH_2O	Pa	filter Δp: 70 Pa; Earth atm.: 100 kPa; tire: 200 kPa; BMEP: 1.2 MPa; diesel injec.: 50 MPa
Specific gravity	See Density: relative			
Specific gravity force	See Density: gravity force			
Specific weight	See Mass or Density: gravity force			
Speed	See Frequency or Velocity			
Stiffness	See Elasticity			
Stress	strength of materials	psi	Pa	strength of steel - tensile: 900 MPa, bending: 200 MPa (numerically MPa equals N/mm^2)
Temperature	fever, weather	°F	K [°C]	human body: 37 °C
	engineering	°F, °R	K	nitrogen liquefies at 77 K; ice melts at 273 K; steel is hot forged at 1300 K
Tension: surface	wetting, droplet formation	lb/ft	N/m	water: 0.05 N/m; ammonia: 0.01 N/m
Time	clock, events of daily life	h, min, sec	s [h, min]	60 min. in 1 h; 24 h in 1 day
	flow measurement, elapsed time	h, min, sec	s	1 ks is 17 min; 100 ks is approx. 1 day; 1 revolution takes 20 ms at 50 rev./s
Torque	See Moment of force			
Vacuum	See Pressure			100% vacuum: 0 Pa absolute pressure; 50% vacuum: 50 kPa pressure differential
Velocity: angular	See Frequency: angular			
longitudinal	vehicle, sound	mile/h, knot, in/sec	m/s	sprinter: 10 m/s; sound in air: 333 m/s, in water: 1444 m/s; light: 300 Mm/s
peripheral	surface speed of grinding wheel	ft/min	m/s	grinding speed: 50 m/s
Viscosity: kinematic	lubrication	centistokes	m^2/s	water: 1 mm^2/s; fuel oil: 5 mm^2/s at 293 K; lub. oil: 10 mm^2/s at 373 K
Volume	capacity, displacement	qt, gal, in^3, ft^3	m^3	beverage can: 1/3 dm^3; baby spoon: 1 cm^3
Weight	See Force or Mass			
Work	See Energy			

*Design dimensions are always in mm in engineering, and the mm is usually not inscribed.

Table 2. SI Equivalents

This is a table of conversion factors, arranged alphabetically according to the respective physical quantities as listed in Table 1. The conversion factors are rounded off to suffice common engineering accuracy.

Examples of conversions among units are shown at the end of the table.

Quantity		Factor	SI unit
Acceleration: angular			
1 degree/sec^2	is	0.0175	rad/s^2
Acceleration: longitudinal			
1 ft/sec^2	is	0.305	m/s^2
1 in/sec^2	is	0.0254	m/s^2
1 g	is	9.81	m/s^2
Angle			
1 degree	is	0.0175	rad
1 min	is	0.291	mrad
1 sec	is	4.84	μrad
Area			
1 acre	is	4047	m^2
1 ft^2	is	0.093	m^2
1 hectare	is	0.01	km^2
1 in^2	is	645	mm^2
1 yard^2	is	0.836	m^2
Coefficient of thermal expansion			
1/°F	is		1.8/K
Conductivity			
1 Btu-in/h-ft^2-°F	is	0.144	W/m·K
Consumption: fuel, specific			
1 lb$_m$/hp (US)-h	is	169	g/MJ
1 g/kW-h	is	0.278	g/MJ
Density: gravity force			
1 lb$_f$/ft^3	is	157	N/m^3
1 lb$_f$/in^3	is	271	kN/m^3
1 kg$_f$/dm^3*	is	9.81	kN/m^3
Modulus of: section			
1 in^3	is	16.4	cm^3
Moment of: area, 1st			
1 in^3	is	16.4	cm^3
Moment of: area, 2nd			
1 in^4	is	41.6	cm^4
Moment of: force (incl. torque and couple)			
1 ft-lb$_f$	is	1.36	N·m
1 in-lb$_f$	is	0.113	N·m
1 in-oz$_f$	is	7.06	mN·m
1 kg$_f$-m*	is	9.81	N·m
Moment of: inertia (incl. Flywheel Effect)			
1 lb$_f$-in-sec^2	is	0.113	kg·m^2
1 lb$_f$-ft-sec^2	is	1.36	kg·m^2
1 kg$_f$-cm-sec^2*	is	0.0981	kg·m^2
1 GD^2 (kg-m^2)	is	0.25	kg·m^2
1 GD^2 (kg-m-sec^2)	is	0.0255	kg·m^2
1 WD^2 (lb-ft^2)	is	0.168	kg·m^2
1 WD^2 (lb-in^2)	is	0.00117	kg·m^2
1 WR^2 (lb-ft^2)	is	0.0421	kg·m^2
1 WR^2 (lb-in^2)	is	0.00029	kg·m^2
Moment of: mass			
1 oz$_m$-ft	is	8.65	g·m
Momentum: angular			
1 slug-ft^2/sec	is	1.36	kg·m^2/s
Momentum: longitudinal			
1 lb$_m$-ft/sec	is	0.138	kg·m/s
1 lb$_m$-in/sec	is	0.0115	kg·m/s

Density: mass			
1 lb_m/ft^3	is	16.02	kg/m^3
1 lb_m/in^3	is	27.68	Mg/m^3
1lb_m/gal (US)	is	119.8	kg/m^3
1 kg/dm^3 or g/cm^3	is	1000	kg/m^3

Density: power			
1 Btu/ft^2-sec	is	11.3	kW/m^2
1 cal/cm^2-sec	is	41.8	kW/m^2
1 W/in^2	is	1.55	kW/m^2

Density: relative

No conversion is usually required for common engineering accuracy.

Dimension (is always in mm in engineering)		
1 ft	is	305
1 in	is	25.4
1 yard	is	914

Elasticity: angular (torsional)			
× deg/100 ft-lb_f	is	7.8/×	kN·m/rad
1 kg_f-m/rad*	is	9.81	N·m/rad
1 lb_f-ft/rad	is	1.36	N·m/rad
1 lb_f-in/rad	is	0.113	N·m/rad

Elasticity: longitudinal			
1 lb_f/ft	is	14.6	N/m
1 lb_f/in	is	175	N/m

Energy: content			
1 Btu/lb_m	is	2.33	kJ/kg
1 cal/g or kcal/kg	is	4.19	kJ/kg

Energy: general			
1 Btu	is	1.055	kJ
1 cal (thermochemical)	is	4.19	kJ
1 Cal (usage in nutr.)	is	4.19	kJ
1 ft-lb	is	1.36	J
1 kg_f-m*	is	9.81	J
1 kW-h	is	3.6	MJ

Energy: specific			
1 Btu/lb_m-°F	is	4.19	kJ/kg·K

Flow: mass			
1 lb_m/min	is	7.56	g/s
1 lb_m/sec	is	0.454	kg/s

Power (incl. heat rate)			
1 Btu/h	is	0.293	W
1 ft-lb_f/min	is	0.0226	W
1 hp (US, mech. eng.)	is	0.746	kW
1 hp (metric)	is	0.735	kW
1 ton (refrigeration)	is	3.52	kW

Pressure (incl. stress)			
1 atm (international)	is	101.3	kPa
1 bar	is	100.0	kPa
1 kg^*/cm^2	is	98.1	kPa
1 lb_f/ft^2	is	47.9	Pa
1 lb_f/in^2	is	6.89	kPa
1 inHg (60°F)	is	3.38	kPa
1 inH_2O (60°F)	is	0.249	kPa
1 mmHg (16°C)	is	133	Pa
1 mmH_2O (16°C)	is	9.80	Pa

Specific gravity - See Density: relative

Specific gravity force - See Density: gravity force

Specific mass - See Density: mass

Specific weight - See Density: gravity force when meaning gravity gravity force per volume, see Density: mass when meaning mass per volume

Speed - See Frequency or Velocity

Stiffness - See Elasticity

Stress - See Pressure

Temperature			
1 °F = 1 °R (increment)	is	0.556	K
1 °F = 461 °R (scale)	is	-16.8 °C	≡ 256.4 K

$t_K = (t_F + 460)/1.8$

$t_K = 5t_R/9$

Tension: surface			
1 lb_f/ft	is	14.6	N/m

Torque - See Moment of force

Velocity: angular - See Frequency: angular

Flow: volume			
1 ft^3/min	is	0.472	dm^3/s
1 gal (US)/h	is	1.05	cm^3/s
1 gal (US)/min	is	63	cm^3/s

Force (including gravity force)		1 knot (international)	
1 dyne	is	0.01	mN
1 kg_f*	is	9.81	N
1 oz_f	is	0.278	N
1 lb_f	is	4.45	N
1 ton_f (short)	is	8.90	kN
1 ton_f (metric)	is	9.81	kN

Frequency: angular			
1 deg/min	is	0.29	mrad/s
1 deg/sec	is	0.0175	rad/s

Frequency: cycle			
1 c/min	is	1/60	Hz

Frequency: rotational			
1 r/min	is	1/60	1/s

Heat - See Energy

Length (see also Dimension)			
1 ft	is	0.305	m
1 in	is	25.4	mm
1 mile (nautical)	is	1.85	km
1 mile (statute)	is	1.61	km
1 yard	is	0.91	m

Mass			
1 carat (metric)	is	0.2	g
1 oz_m (avoirdupois)	is	28.35	g
1 oz_m (troy)	is	31.10	g
1 lb_m	is	0.454	kg
1 slug	is	14.6	kg
1 ton_m (short)	is	0.907	Mg

Modulus of: elasticity			
1 lb_f/in^2	is	6.89	kPa

Velocity: longitudinal, peripheral			
1 ft/min	is	0.0051	m/s
1 ft/sec	is	0.305	m/s
1 in/sec	is	0.0254	m/s
1 km/h	is	0.278	m/s
is	0.515	m/s	
1 mile (statute)/h	is	0.447	m/s

Viscosity			
1 centipoise	is	1	mPa·s
1 centistokes	is	1	mm^2/s

Volume (incl. capacity)			
1 barrel (US, liq. exc. oil)	is	0.12	m^3
1 barrel (oil)	is	0.16	m^3
1 ft^3	is	0.028	m^3
1 gal (US)	is	3.79	dm^3
1 in^3	is	16.4	cm^3
1 oz (US, liquid)	is	29.6	cm^3
1 quart (dry)	is	1.101	dm^3
1 quart (US, liquid)	is	0.946	dm^3
1 $yard^3$	is	0.765	m^3

Weight - See Force when meaning gravity force, see Mass when meaning mass

Work - See Energy

Use of this Table:

1) Converting to SI units: Multiply the known value by the equivalent. Example:
   Flow of 15 lb_m/min. How much is it in g/s?
   Since ... 1 lb_m/min is ... 7.56 g/s,
   therefore ... 15 lb_m/min is ... $15 \times 7.56 = 113.4$ g/s

2) Converting to in-lb units: Divide the known value by the equivalent. Example:
   Flow of 15 g/s. How much is it in lb_m/min?
   Since ... 1 lb_m/min is ... 7.56 g/s,
   therefore ... 15 g/s is ... $1/7.56 = 1.98$ lb_m/min

* kg_f = kg force, sometimes written as kp or kg*

Appendix A

Basic IC Engine Performance Formulae

Calculating Test Data in SI Units

Power

Since power = torque × angular velocity,

$$P = T \times 2\pi \times RPS$$

Example (engine of 1000 N·m at 40 s^{-1}):

$$P = 2\pi \times 1 \text{ kN·m} \times 40 \text{ s}^{-1} = 250 \text{ kW}$$

Mean Effective Pressure

Since MEP × piston area × stroke = work of one cylinder per cycle,

$$MEP = \frac{2\pi \times T}{V \times Z} \text{ or } \frac{P}{V \times Z \times RPS}$$

where V is the displacement volume of one cylinder, or trochoid, and Z is the number of all firing strokes in one revolution (i.e., number of cylinders with two-stroke engines, number of rotors with wankel engines, number of cylinders-divided-by-two with four-stroke engines).

Example (four-stroke, six-cylinder engine of 125 mm bore and 130 mm stroke, at 770 N·m torque):

$$BMEP = \frac{8 \times 770 \text{ N·m}}{0.125^2 \text{ m}^2 \times 0.130 \text{ m} \times 3} = 1 \text{ MPa}$$

Specific Fuel Consumption

Since specific fuel consumption = amount of fuel per unit work,

$$\text{SFC} = \frac{\text{mass fuel flow}}{\text{power}}$$

Example (647 g measured in 100 s, at 90 kW):

$$\text{BSFC} = \frac{6.47 \text{ g/s}}{90 \text{ kW}} = 72 \text{ g/MJ}$$

Specific Energy Consumption

Since specific energy consumption = amount of input energy per unit work,

$$\text{SEC} = \text{SFC} \times \text{lower heating value}$$

Example (the above engine, using 41.8 MJ/kg fuel):

$$\text{BSFC} = 0.072 \text{ kg/MJ} \times 41.8 \text{ MJ/kg} = 3.00$$

Cycle Fuel Delivery

Since fuel volume per cycle = mass per intake divided by density,

$$\text{FVPC} = \frac{\text{mass flow}}{\text{density} \times Z \times \text{RPS}} \text{ or } \frac{\text{volumetric flow}}{Z \times \text{RPS}}$$

Example (the above fuel flow and engine, fuel density of 837 kg/m^3):

$$\text{FVPC} = \frac{0.00647 \text{ kg/s}}{837 \text{ kg/m}^3 \times 3 \times 40 \text{ s}^{-1}} = 64 \text{ mm}^3\text{/intake}$$

Appendix B

Background Information on Metrication in America

This appendix contains excerpts from the 1975 and 1988 metric legislations, a brief history of metrication in the U.S., some experiences of those who metricated, tips on the adoption of international standards and practices, and tips on how to become proficient and comfortable working with metric units.

Metrication - Why and What is the Status

Global technology, global commerce, and global markets are becoming a reality as this century closes. The globality requires that products be designed for universal acceptance in manufacturing and selling. The implementation of this strategy is unthinkable without a unified, global system of measurement units.

Throughout the history of civilization, attempts have been made to limit the number of the measurement systems in use, and, ultimately, to agree on one. The attempts narrowed the field eventually to only two systems, the so-called English or inch-pound system, and metric. Globally, the metric system has been gaining ground while the inch-pound system has been losing. Today, only the United States (and perhaps Myanmar and Liberia) remain uncommitted to changeover in the sense that no nationwide mandate for metrication has been established.

Metric and America

The system of units to which the world is converging is not just a "metric system." There are different "metric systems," or, better to say, versions of it. They were established at various points in the two centuries of the system's evolution, and the major ones were named, for example, CGS, MKS, and others.

The latest major revision, in 1960, coined the present name, The International System of Units, or SI; this is the way the system should be referred to in order to avoid ambiguity. The letters SI have been adopted by all nations and languages of the world to identify the new system.

Note: When the term "metric" is used in this text, it is for brevity when referring to no particular version of the system.

SI was adopted by the International Organization for Standardization (ISO) in 1973, and reached worldwide recognition in the '80s. Most professionals use only SI units in new designs and projects today, and the younger generation is educated in SI exclusively in all nations except the U.S.

It should be stressed that the changeover to metric in the U.S. means a changeover to SI. This is by the decree of all the professional and educational societies that took an active part in this decision making; the Government has taken scant interest in any of the metric affairs.

Historically, the lack of a resolute involvement of the Government in establishing a simpler measurement system in this country is surprising in light of the importance our founding fathers placed on simplicity in other areas of public interest. Thanks to Thomas Jefferson, for instance, our nation was first to have a decimal coinage plan (ten mills to a cent, ten cents to a dime, ten dimes to a dollar). He also proposed simplifications in other areas of measures, such as dividing the day into decimal increments, and eventually also adopting the metric system.

Several times since Thomas Jefferson's attempts, the Congress toyed with the idea of adopting the metric system, and scored some adjustments, the major ones listed below. Until recently, the most prevalent argument against the adoption involved the U.S. dependence on the trade with the British Commonwealth of Nations.

— In 1866 a compromise of sorts had been obtained by an Act of Congress to legalize the use of metric measures.

— In 1893 the U.S. became an officially metric nation when all standard measures such as yard, inch and pound were redefined to become a multiple of the corresponding metric measures.

— In 1902 Congress voted on a bill that would have required all government departments to use the metric system exclusively. The bill was defeated by one vote.

Overall, several dozen attempts for metrication have been registered in the last hundred years in national legislation, none of which received much notice from the American public.

— In 1975, however, a law called the Metric Conversion Act of 1975 was passed and this one caught public attention. America started preparing for a metric future. Concurrently, all the remaining non-metric countries, which were the English-speaking countries and a host of smaller nations, made commitments to abandon their respective systems and adopt SI.

A decade later, all of those countries with the exception of the U.S. had accomplished the task. Why did the changeover fail in the U.S.?

This question cannot be answered here for two reasons:
(1) the elaboration would be too long for the purpose of this treatise, and
(2) it is not universally accepted that the changeover failed.

However, a brief summary of the metrication developments in the decade since the Metric Conversion Act of 1975 was enacted is presented.

Recent Metrication Developments

The 1975 law called for a voluntary conversion. Most consumer-oriented U.S. companies did not change their product; a voluntary conversion was perceived by them as possible annoyance to their customer, the general public.

On the other hand, metrication has taken hold in the global and export-oriented companies, particularly in the automotive and agriculture industries. For example, the design work on new automobiles was changed to metric by the late '70s; the new cars reached the showrooms some five years later. The wine and spirits industries changed to solely metric packaging in the early '80s.

Schools started teaching metric in the '70s, but relaxed it later, responding to the apathy of the American public toward the adoption of the new measuring system.

On the surface, it seemed metrication was losing ground rather than gaining it with time. In reality, however, a great deal of work had been done in the private industry. And while the American public was relaxing on the issue, all the remaining non-metric countries (Australia, Great Britain, South Africa, Canada, ...) adopted SI.

Recognizing the nation's isolation and yielding to the urging of NATO, commercial allies, and many U.S. citizens and groups, Congress was moved to modify the 1975 law.

— In 1988, President Reagan signed into law the Omnibus Trade and Competitiveness Act. A section in this document, P. L. 100-418 designates the metric system as the preferred system of measurement for the Federal Government. It amends the Metric Conversion Act of 1975 by stating:

"SEC. 3. It is therefore the declared policy of the United States

"(1) to designate the metric system of measurement as the preferred system of weights and measures for United States trade and commerce;

"(2) to require that each Federal Agency, by a date certain and to the extent economically feasible by the end of the fiscal year 1992, use the metric system of measurement in its procurements, grants, and other business-related activities, except to the extent that such use is impractical or is likely to cause significant inefficiencies or loss of markets to United States firms, such as when foreign competitors are producing competing products in non-metric units;

"(3) to seek out ways to increase understanding of the metric system of measurement through educational information and guidance and in Government publications, ..."

Even prior to the passage of this public law, the awareness of the need for the metric changeover grew among the Federal agencies. For example, the DOD stipulated in a September 16, 1987 directive to:

> "use the metric system in all of its activities, consistent with security, operational, economical, technical, logistical, and safety requirements."

The DOD was more specific in the directive issued for various projects. For example, the Strategic Defense Initiative Organization issued a metric policy which stated:

> "All newly designed, developed and produced systems and elements that make up the Strategic Defense System (SDS) shall use SI metric units as the standard language and system of measurement."

In aerospace, the LHX helicopter program is metric. A study predicted 0.5% higher developmental cost, which would be recovered in lower manufacturing cost after a dozen or so helicopters are built. The actual cost increase was found to be about half the prediction, making the prospects for the quick cost recovery bright indeed.

The military also has a project with the Society of Automotive Engineers (SAE) to rewrite military (MIL) specifications and standards to SI. Concurrently it has made a commitment to use industrial standards wherever they can replace the MIL ones, and to issue all new standards in SI.

Similarly to DOD, NASA has made a commitment to run all new programs using SI units. Even the Space Station, for example, which was a metric program later redirected to the use of inch-pound units as a consequence of the safety concerns after the space shuttle Challenger accident, is expected to revert back to metric.

— Finally, on July 25, 1991, President Bush signed an Executive Order directing the Commerce Secretary to coordinate the effort of the Federal Agencies in complying with the law.

There can be no doubt that the metric system is the system of international commerce. There is also no doubt that metrication has progressed much further in the U.S. than is generally believed.

The next pages present some of the issues to be considered in the changeover to metric. First we shall explore international standardization, and how standards develop.

Standards: ISO and Others

An industrial or commercial standard is a document which describes the size, shape, color, function, and other characteristics of an object or process. Standards serve the needs of a wide range of human endeavors, and there are a great many of them. They enable hair dryer plugs to fit their sockets; they prescribe the air pollution limits, composition of a steel, voting procedure, etc.; the range seems endless. If it were not for standards, living would be a lot more complicated and expensive.

Evolution of Standards

Most larger organizations everywhere in the world have been writing standards to enable their business to function more effectively. Some of these standards have wider appeal and get adopted by other organizations. Eventually a standard may become nationwide, later regional, and ultimately international (global).

Today all industrial nations have their standards and standards-setting institution(s). In the U.S. there are over four hundred groups that issue standards: ANSI, ASME, IEEE, ASTM, SAE, to name a few. Most other nations have just one such institution; standards of these nations are referred to by the institution's abbreviation or acronym. For example, JIS for industrial standards of Japan, AFNOR for France, BSI for Great Britain, GOST for the (former) Soviet Union, DIN for Germany.

Among the national standards, the DIN (Deutsche Institut fur Normung) standards are the best known ones. The extent of the popularity of DIN standards is such that in the U.S., for example, DIN nuts and bolts are often considered synonymous with metric nuts and bolts; this assumption is, of course, not completely correct.

Among the regional standards, the European Community (EC) standards will become the most important ones. The EC standards are known as:
— the European Committee for Standardization or CEN standards, and
— the European Committee for Electrotechnical Standardization or CENELEC standards.

The CEN/CENELEC standards are issued for the most populous industrialized market in the world. Furthermore, these standards are being adopted by all European and formerly Communist Bloc countries, and by the whole world, for that matter, that use metric-units-based standards.

On a global level, and for engineering purposes, the major organization responsible for international standards is the International Organization for Standardization (ISO) and the International Electrotechnical Commission (IEC). Almost one hundred countries are members of ISO.

A standard becomes an ISO or IEC standard after a lengthy process of international reviews and approvals by the member nations. Although the vast majority of these standards use metric units, there are also non-metric standards in the ISO/IEC. ISO does not necessarily mean metric, while, on the other hand, going metric usually means going ISO.

Harmonization and Standardization

Harmonization is the standardization of standards, practices, procedures and tariffs internationally. Nowadays, it is progressing on all fronts. For example, the U.S. and about 30 other nations have agreed to use the Harmonized Tariff System (HTS) in international commerce. The HTS simplifies the documentation that accompanies nation-to-nation shipments, it standardizes procedures, mandates the using of only SI units, and reduces the number of similar standards.

There are several hundred thousand standards in existence worldwide. Many are copies of each other, many exhibit insignificant differences, many exhibit substantial differences but are no longer used. In the U.S. alone, there are close to 100 000 nationally recognized standards. Although the number of topics in need of standardization grows with the technological progress of our civilization, the number of different standards written for the same item will narrow down, until, hopefully, just one standard will cover that need for everyone.

The ISO has issued by now almost 10 000 standards. The trend among most nations and standards-setting institutions is to adopt ISO/IEC standards where they exist, to work through worldwide organizations on improving existing standards, and seek incorporation of other successful standards into an international body.

It is the explicit policy of the EC, for instance, to incorporate ISO/IEC standards wherever they exist. If a needed standard is not available from that source, only then is a new one written, or an existing, perhaps a national one, adopted. Similar policies are now being issued by several U.S. standards-writing bodies. In the case of the "MIL specs," for example, a policy requires their replacement with equivalent industrial standards.

Harmonization and SI

The world, with the exception of some U.S. industries, has been harmonizing on ISO standards for decades. One of the most basic ISO standards is the ISO 1000 document, which is the standard describing SI. It has become the sole standard for units and the measurement language worldwide. It is expected that when the U.S. adopts the metric system it will be in harmony with ISO 1000.

Standards reflect the needs of civilization and the evolution of technology. Hundreds of new standards are written yearly, and all existing standards are reviewed and improved continuously, typically in five-year increments.

The standard ISO 1000 is also developing. Few changes are expected, however, in the commonly encountered units for decades to come. The thrust of the work is in the development of units for new scientific disciplines.

Which way is the best way to adopt international engineering practices? Some of these issues are explored next.

Changing Over to Metric

The slow progress of metrication in this country necessitates that engineering professionals work using two systems of measurement — inch-pound and SI. They need to know two systems because their employers are working on inch-designed products alongside metric ones. To be productive in this environment, employees need to be fluent in both systems.

Such fluency does not come from memorizing conversion factors and performing calculations. On the contrary, converting delays the learning process. Converting also lowers productivity by increasing the time the process takes and by possibly introducing errors. Studies show that an 8% error rate is usual in performing conversions. Converting is a U.S. handicap; other nations are not burdened with it anymore.

Changing over should mean phasing-out the non-international practices, not converting.

Avoid Converting Units

There is little need for converting units if you can "think metric" and have the feel for the sizes of SI units. The material presented in this book accompanied by appropriate training will, in most cases, be enough for you to acquire the ability and feel with practice. "Having the feel" will make you comfortable working in SI and avoid low-level errors.

You will also work more efficiently if the SI figures feel in the "ballpark." Acquiring that feel takes some effort, but without it, engineers can waste a lot of time checking and rechecking data.

Avoid Converting Designs

Converting designs to metric, rather than drafting them according to the metric engineering and drafting practice from the beginning, means higher manufacturing and quality control cost. There is little compatibility between metric and inch-pound standard hardware. Each is based on its own rules for standardizing sizes. Therefore, to be cost effective in the world market, most designs must be executed from inception as either metric (international) or inch-pound projects. Designers should not rely on the "in/mm" switch on their CAD systems to produce metric drawings. Many know that, but must resort to such practice when their departments regard this switching capability as a way to avoid spending the time on training and on providing metric design and drafting manuals.

Just as with the units, it is important to become "fluent" with the international standards for the often-used items in your discipline. That means developing a feel for sizes, shapes, strength, and so on. For example, an M12 thread should be just as readily recognized as a 1/2"-13 NC thread. And features that make a metric bolt metric should be apparent. This can be done only by making a conscious mental effort during training and in daily encounters. The material presented here will put you on the right track; but the learning must go on.

Changing the "Soft" or "Hard" Way?

The terms "soft metric" (or "soft conversion") and "hard metric" (or, incorrectly, "hard conversion") are used to describe two approaches to the metric changeover.

SOFT

Changing units, not product

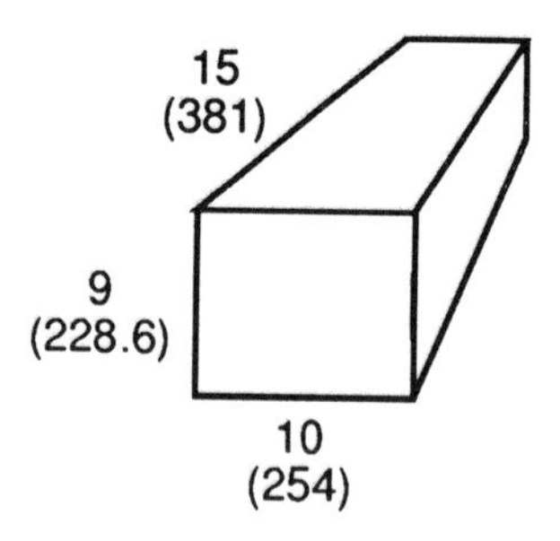

HARD

Producing metric design

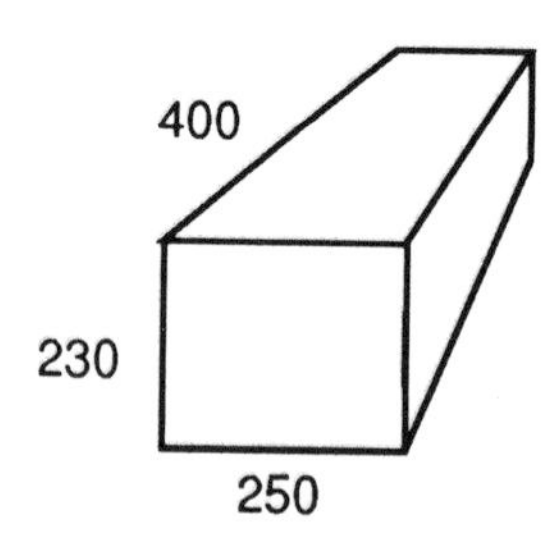

Soft metric implies a conversion by calculation. The metric size and specification numbers are simply calculated from the in-lb units (and, hopefully, rounded to reflect implied accuracy as was presented in Chapter 6). The resulting numbers are most likely nonstandard in the metric world. The product's size and dimensions are the same as before.

Although the soft conversion should be an exception rather than the rule for the changeover, it has its place and had been practiced for generations. Most railroads worldwide, for example, have rails 1435 mm apart, that figure being a conversion from the British "gauge" of 4 ft 8-1/2 in. Similarly, the sizes of socket wrench drives and light bulb threads were metricated by soft conversion. Such designs exist in both in-lb as well as metric engineering. An example of a soft conversion to in-lb is the sizing of common rolling bearings. These universally adopted features are referred to as *hybrid* designs.

Hard metric implies a totally metric design. The product's dimension and specification numbers conform to the internationally standardized design features and practices. Typically, a metric design exhibits the following:

- data are presented in SI units
- drawings are made to the ISO drafting practice
- dimensions are in whole numbers of mm
- sizes follow the preferred number series and/or are in use internationally
- applicable standards are in accordance with ISO/IEC documents

What Strategy for Metricating a Company?

There are two basic strategies for companies faced with a changeover.

— In one, the company retains all present practices on existing products and establishes metric practices only with new products. This strategy requires little soft conversion.

— The other strategy means a quick, company-wide changeover. In this case, it is inevitable that most of the existing product documentation is soft converted to keep production going.

The "right" strategy will usually lie somewhere between these two extremes, and should be decided only with careful consideration of engineering, purchasing, marketing, and manufacturing concerns.

Bibliography

A comprehensive bibliography of the more recent metric literature has been made available in one volume titled Freeman Training/Education Metric Material List. The volume was compiled by the U.S. Metric Association, and it is sold through the headquarters of the Association (see address listed on next page). This author refers readers to that source for the bibliography for the following reason.

Even a casual searcher of metric publications soon realizes that the number of books, articles and training aids published in this country on the subject of metric units and metrication is overpowering. To list just the material used for this author's research would amount to a hundred entries, and to list some of the references and not others might provide a disservice by promoting one work over another. Most of the literature presents similar topics, the difference being in the approach and the extent of the topics.

If the searcher's interest lies in units only, he/she is advised to consider that all modern publications dealing with SI units are based on a thin booklet called ISO 1000 (ISO means International Organization for Standardization). This document in turn is based on the work produced by the CGPM (an abbreviation for the committee responsible for the development of SI).

The ISO document is available from a number of commercial sources, and also from the ISO headquarters, 1 Rue De Varembe, Case Postale 56, CH 1211 Geneve 20, Switzerland, or from ANSI (American National Standards Institute), 11 W. 42nd St, 13th Fl., New York, NY 10036, ANSI being the official representative of ISO in this country.

The CGPM work, which traces the development of the metric system through most of its history, is available from the BIPM (an abbreviation for the bureau that houses the unit standards), Pavillon de Breteuil, F-92310 Sevres, France. The American version of the English section of this fundamental book is issued by NIST (National Institute for Standards and Technology, formerly NBS) (see address listed on next page) under the heading Special Publication 330. It is available from the Superintendent of Documents, Government Printing Office, Washington, DC 20402-9325.

Those seeking further assistance are referred to the three non-profit organizations dedicated to providing metrication assistance in this country. Their addresses follow.

Metric Program, Dept. of Commerce
NIST
Bldg 101, Rm. 813
Gaithersburg, MD 20899
Telephone: 301-795-3690

(A Federal Government funded organization)

U.S. Metric Association
10245 Andasol Avenue
Northridge, CA 91325
Telephone: 818-363-5606

American National Metric Council
1735 N. Lynn St., Suite 950
Arlington, VA 22209-2022
Telephone: 202-857-0474

Index

About the Author

Stan Jakuba is president of a consulting firm that specializes in training business leaders, technical personnel, and educators in international standards.

In his professional career, Stan has worked in academia and industry here and in several countries of Europe. He is a graduate of M.I.T., Cambridge, Mass., and C.V.U.T., Prague, Czechoslovakia. Progressing in his employs from a worker, technician, teacher, engineer to manager and professor, he is now a recognized trainer, speaker, and consultant.

Stan has written numerous papers and articles on teaching and practicing SI and on international standards.

He is a member of several standards-setting committees, the ASME Metric Board and Design Education Committee, Vice President of the U.S. Metric Association, a member of the Metric Practice Committee of ASTM, Metric Advisory Committee of the Society of Automotive Engineers, Standards Engineering Society, American National Metric Council, and the metric committee of IEEE.

For over a decade Stan has been helping U.S. companies to "go metric," making it possible for them to adopt international standards and to save.